D0554211

Reconstructing Large-scale Climatic Patterns

from Tree-Ring Data

HAROLD C. FRITTS

Reconstructing Large-scale Climatic Patterns from Tree-Ring Data

A DIAGNOSTIC ANALYSIS

The University of Arizona Press

TUCSON & LONDON

The University of Arizona Press
Copyright © 1991
The Arizona Board of Regents
All Rights Reserved

♾ This book is printed on acid-free, archival-quality paper.
Manufactured in the United States of America.

96   95   94   93   92   91   6   5   4   3   2   1

Library of Congress Cataloging-in-Publication Data
Fritts, Harold C., 1928–
    Reconstructing large-scale climatic patterns from tree-ring data:
a diagnostic analysis / Harold C. Fritts.
        p.   cm.
    Includes index.
    ISBN 0-8165-1218-3
    1. Dendroclimatology — North America.   I. Title.
QC884.2.D4F73   1991
551.697 — dc20   91-15764
                                                                   CIP

British Library Cataloguing-in-Publication Data
A catalogue record for this book is available from the British Library.

Designed by Laury A. Egan

To Barbara and Miriam

Contents

Figures

Tables

Equations

Foreword

I greatly welcome the appearance of this work in print. Professor Hal Fritts has been the leading figure in the second phase of the development of dendroclimatology, the great blossoming of the subject. The key achievement has been its adaptation to the computer age, which has enormously widened the capacity of tree-ring studies to ravel the secrets of climatic history. Such adaptation has been Fritts's special mission, and this book presents the fruition of thirty years of energetic research and the commitment of his ever-active mind to the subject. In the process, Fritts has made contacts in many countries and with leading workers in other disciplines, and he has inspired people everywhere to see how far they can proceed with similar and related work. Thus the prospect of worldwide coordinated surveys has been opened up.

Interest in the possibility of deciphering the messages preserved in tree rings goes back at least to Leonardo da Vinci. Application of this potential to the detection of past climate actually began to be systematically explored in this century, from 1901 onward, by A. E. Douglass, who came from Harvard College's astronomical observatory to Arizona searching for traces of associations between sunspots, tree growth, and weather. Recognition—through the narrow rings introduced into the sequences by particularly stressful drought years—of similar tree-ring sequence patterns in the wood from trees over wide areas soon led to the idea of cross-dating from tree to tree and so to the development of regional chronologies. It was some time before the "floating chronologies" shown by the beams in prehistoric buildings in various parts of the southwestern United States could be linked to the chronologies derived from beams in later buildings and then to those from living trees. But by about 1920 dates could be given for Aztec buildings in New Mexico and then to the Pueblo Bonito prehistoric Indian buildings. Thus the power of dendrochronology was demonstrated.

The arid southwestern sector of the United States was also an ideal region for exploring how this dendrochronology could be applied to interpretation of the past climatic behavior on which it was based. As early as 1911 to 1913, Ellsworth Huntington, measuring the rings in 451

Sequoia trees in the Sierra Nevada in California, went on to look for a possibly parallel course of the "fossil records" from the trees of droughts over the last three thousand years in California and in central and southwest Asia at about the same latitude. Between around 1910 and the 1940s, first Douglass, and then his student E. Schulman, built up a considerable knowledge—unmatched at the time—of the apparent history of droughts and rainfall variations in the American Southwest.

The further advance of dendroclimatology, deriving specific and reliable information about past climate, was not possible until the complex relationships between tree growth and the soil moisture and weather sequence at the site where a tree grows could be analyzed thoroughly. The University of Arizona Laboratory of Tree-Ring Research at Tucson nurtured a group of dedicated scientists, gradually amassing an unmatched wealth of the relevant knowledge and an ever-growing bank of tree-ring measurements. Progress had to wait on the development of the necessary computing capacity and the devising of multivariate analysis techniques for handling the data from very large numbers of carefully selected trees and tree sites. Understanding of tree biology, of cell growth, and of the influences of different kinds of site and soil was continually increasing. Once the computers and the analytical routines were there, the representativeness of individual trees and tree sites, and the effects of various weather sequences over periods of up to fifteen months, began to be explored. Later, another vital step was to use the relationships discovered between tree growth and weather to take the tree-ring data and "back-cast" (i.e., reconstruct) the climate of previous periods within the present century and then test the result by comparison with the known reality.

The logic of the successive testing stages and the progress of the scheme devised by Fritts are recounted with admirable clarity in this work. Verification tests were applied at every stage, and their validity is discussed. It is a great story.

The whole is an outstandingly careful work that makes no false claims. But the outcome is an exciting one. It provides in these pages a compendium of climatic history from 1602 to the present century which, particularly over the western half of North America, could hardly have been obtained by any other means. This is a marvelous contribution to paleoclimatic study, a firm basis for comparisons with what can be reconstructed of the contemporary climatic patterns over Europe and other parts of the Northern Hemisphere from documented observations and other sources. As it turns out from Fritts's tests of error liabilities, the temperature reconstructions derived from the tree-ring data are more reliable than the precipitation values indicated. Annual values and averages over several years are more reliable than seasonal or shorter-period values. And not surprisingly, in view of where the trees were, the results for western North America are more reliable than those for places farther east. That, of course, is a great virtue, because it is precisely for western North America that we have the greatest lack of alternative types of detailed record. One fascinating indication is that most of western North America was somewhat warmer than now during the Little Ice Age in the seventeenth to nineteenth centuries, when the rest of North America shared the experience of Europe (and of much of the world) in being colder than now. Such anomaly characteristics are described and discussed: they seem to imply some further aspects of the patterns far outside North America so that, although great care must always be taken about limits of reliability, the climatic maps derived from these studies seem to have lessons to teach us about the contemporary climatic regime over much of the hemisphere.

We are living in a time when much attention and many reasonable anxieties are focused on the manifold threats to the living world about us, and all too many people are inclined to think that we don't need any more studies of history: "Throw the records of the past out the window!" This reaction is to forget that the climate is, and always has been, changing this way and that, to greater or lesser degree, under the impact of natural causes—solar fluctuations both short and long term, volcanic explosions, and so on. And tree-ring data provide a great fund of material for exploring the workings of these influences in the past. It is still an open question—and one that needs to be answered—how far current climatic changes are due to natural causes and long-term fluctuations that can be traced far back into the past. While the latest scientific literature and an enormous amount of popular writing are absorbed with the expectation of warmer climate because of the greenhouse effect of increasing carbon dioxide and other gases, it is rarely remembered that somewhat greater warmth than in the period from 1900 to 1950 in the tropical Atlantic and in latitudes near the Antarctic was also indicated by observations in the Little Ice Age, when the higher northern latitudes were generally colder, as they have been becoming since about 1960.

Just how these latest anomalies should be understood is not yet established, but it is clear that we still greatly need all possible sources of information on the climates of the past. Tree-ring data carefully handled are, as this work shows, a very rich resource, one that provides some detail and can be extended to other parts of the world and to data from still earlier times.

Hubert H. Lamb

Preface

In 1960 I joined the University of Arizona's Laboratory of Tree-Ring Research in the faculty position previously held by Edmund Schulman. As a graduate student, I had been fascinated with Waldo S. Glock's public criticisms of Schulman and his associates and had decided to investigate these objections carefully. My work began with a systematic analysis of tree-ring features from the point of view of a skeptical botanist. I constructed dendrochronological hypotheses, sampled dendrochronological data along ecological gradients, applied objective statistical analyses, and tested the hypotheses in the context of known biological processes.

It became apparent that many of Glock's criticisms must have arisen out of personal differences between Glock and Schulman, not from real scientific questions, for the scientific principles espoused by both Schulman and Douglass were generally confirmed by my analysis. Needless to say, as an objective scientist I could only conclude that the Douglass system was valid. Cross-dating worked, site selection was important for signal strength enhancement, and standardization was important to transform ring-width measurement to a stationary time series.

Much of my early work involved restatements of Douglass's and Schulman's ideas from a biological perspective. This work was described in journal articles and has since been summarized in *Tree Rings and Climate* (Fritts 1976). As I reflect on this first decade with the laboratory, I realize how much I drew intuitively from the ideas of my predecessors, although at the time I was convinced that they were entirely my own experiments and analysis, which they were on the surface. Now I wish to acknowledge my profound inheritance from both A. E. Douglass and Edmund Schulman. Other important influences in my early dendrochronological upbringing were T. L. Smiley, W. G. McGinnis, B. Bannister, M. A. Stokes, and C. W. Ferguson. They were extremely tolerant of my independence, receptive to new approaches, and accepting of my limitations (for this young botanist would never become proficient in cross-dating procedures). Later a young geologist,

V. C. LaMarche, Jr., joined the laboratory and became a trusted colleague whom I depended on to test new ideas.

During the second decade of my tenure with the Laboratory of Tree-Ring Research I focused on a lifetime dream to use grids of tree-ring data to reconstruct grids of climatic data. The beginnings of this work were outlined in the last chapters of *Tree Rings and Climate*, but the scientific effort really had begun early in the 1970s with a systematic recollection of arid-site tree-ring chronologies to update the Schulman (1956) analysis. The collection, dating, and analysis of the basic tree-ring data would not have been possible if it were not for my colleagues W. G. McGinnis, M. A. Stokes, T. P. Harlan, C. W. Ferguson, M. L. Parker, W. J. Robinson, J. S. Dean, and R. A. Holmes, who were largely responsible for assembling the first dendrochronological networks.

This book describes this effort, which culminates with reconstructed seasonal and annual temperature, precipitation, and sea-level pressure. Four individuals played important leadership roles in this work: T. J. Blasing, G. A. Gordon, G. R. Lofgren, and J. M. Lough. T. J. Blasing was instrumental in establishing the program and research strategy. Gordon, Lofgren, and I initially signed a contract with Academic Press for this volume. However, the task turned out to be larger than we anticipated; Gordon went on to more challenging opportunities; Lofgren assumed responsibility for the data processing at the laboratory, and Academic Press changed management and policy, which resulted in a broken contract. Gordon wrote two chapters on spectral analysis of seasonal reconstructions, but the sheer size of the analysis required us to focus primarily on the annual data. By the time I had pulled together the first draft of this manuscript, too much time had elapsed, and Gordon and I did not manage to finish the collaboration. I have tried to do justice to his good work, and I take complete responsibility for any errors in the writing.

J. M. Lough and I managed to publish much but not all of our collaborative efforts. We had invested considerable time in compiling files on various proxy evidence for climatic variation, and we had hoped to use these files more fully in analyzing and testing the reconstructions. Funding was terminated before we completed this activity, and it was necessary to abandon the work until this book was finished. Now climatic reconstruction work has passed into other hands, and I am focusing more of my efforts in the area of dendrochronological modeling.

Many details involving the seasonal climatic reconstructions have not been included here. Maps have been made for all seasons and all years. An analysis of variance and covariance of seasonal, annual, and decadal variations was finished but included too much detail for this volume.

Power spectra were calculated and time-series plots were constructed for all seasons and regions. The transfer-function coefficients were rearranged in terms of the original tree-ring chronologies, and detailed maps of these coefficients have been produced so that one can evaluate the transfer-function relationships that were evident. All of the seasonal reconstructions are available for anyone wishing to have them. I have no plans for publishing another volume. Some of the work is described in unpublished technical reports of the Laboratory of Tree-Ring Research, but too much remains in files at my office.

Acknowledgments

I acknowledge the significant contributions of T. J. Blasing, G. A. Gordon, G. R. Lofgren, J. M. Lough, and R. L. Holmes. Blasing was largely responsible for the development of the canonical regression analysis; Gordon developed the array of verification tests; and Lofgren managed the data base and technical staff working on the project. Lough used the reconstructions to examine some important climatic issues and questions. Holmes led new collection efforts and contributed significantly to certain technical aspects of the project. Programmers included J. H. Hunt, K. C. Sakai, B. J. Keller, M. Perez-Wilson, and J. G. Rapp. M. A. Burgess, L. E. Conkey, J. P. Cropper, C. J. Earle, Shao Xuemei, D. N. Duvick, H. Cathy, K. K. Hirschboeck, M. Thompson, and L. C. Winter were graduate students at some time, and many became full-time workers in this field after finishing their academic requirements. J. M. Burnes, J. R. Carter, J. Harsha, E. DeWitt, D. J. Shatz, and D. W. Stevens served in various technical capacities. M. Huggins, M. Harington, D. A. Larson, J. Mather, A. Marek, B. J. Molloy, K. B. McDougall, N. Noble, and J. A. Sherwood helped in the preparation of the manuscript. J. Johnson, T. Paczosa, N. Pranke, K. C. Barber, J. Wagner, A. Woodley, M. A. Ytuarte, M. Mague, M. Carey, and R. T. Will were employed on the project. C. W. Stockton, L. G. Drew, and Wu Xiangding contributed informally to the project. I also acknowledge the support of my late wife, Barbara; my children, Marcia and Paul; and my present wife, Miriam, who contributed in numerous ways to the success of the project.

This material is based on work supported by the National Science Foundation under Grant Nos. GA 26581, ATM 75-22378, ATM 77-19216, ATM 8115754, and ATM 8319848. Any opinions, findings, conclusions, or recommendations expressed in this publication are my own and do not necessarily reflect the views of the University of Arizona or the National Science Foundation.

Reconstructing Large-scale Climatic Patterns

from Tree-Ring Data

1

Introduction

A. The Importance of Climatic Variability

The earth has undergone significant variations and changes in climate over a variety of time scales, and we have every reason to expect that such variations will continue (National Academy of Sciences 1975). In recent years people have become increasingly concerned with the possibility that climate could be undergoing a change or that human activities might tip the balance of the climate system and bring about unanticipated climatic changes (Kellogg 1987; Wood 1988b).

A well-documented history of past climatic conditions is needed to understand the causes of the variation and to differentiate between natural variation and inadvertent anthropogenic changes. In addition, understanding the history of climatic variation could lead to a more fundamental appreciation of our earth, its biota, and the development of civilization.

Various simulation models have been developed to study the causes of climatic variation and change and to help humankind anticipate future conditions (Hecht 1985; Schlesinger and Mitchell 1987; Wood 1988b). The global climate system is immensely complex, however, and models of that system are based on many assumptions that could lead to large uncertainties in the model results (Saltzman 1985; Dickinson 1989; Wood 1988b). For example, many models are based on relationships observed in the instrumental data that have been collected mostly from the twentieth century. One critical assumption of these models is that they express well-known laws of energy conservation and can make reliable projections of future conditions even beyond those experienced in the twentieth century. A well-reconstructed history of past climatic variations over the earth could extend our knowledge to time periods of different climatic conditions, which would help to test such assumptions.

In addition, climate is a time-transgressive phenomenon, so models for assessing possible climatic changes must consider the factors that governed the shorter-term historical climatic variations and intermediate-term changes, as well as the longer-term variations (Hecht 1985). For example, climate models of orbital and plate-tectonic forcing were conceived largely from the strong paleoclimatic evidence for that forcing but included the basic energy-conservation relationships drawn primarily from the twentieth-century instrumental record (Barron et al. 1981; Barron, Thompson, and Schneider 1981; Imbrie and Imbrie 1980; Hecht 1985). Also, the accuracy of modeling the short-term climatic effects of large explosive volcanic eruptions or the influence of El Niño/Southern Oscillation (ENSO) events depends, in part, on the accumulation of accurate histories of climatic variations and related events. The twentieth-century climatic record spans very few major volcanic eruptions, so it is necessary to examine the climate in prior centuries to evaluate the full extent to which volcanic dust veils influence climate.

B. The Role of Paleoclimatology

Paleoclimatic variation encompasses a wide range of time scales, and the associated causes of climatic change operate at different frequencies. Much paleoclimatic work has focused on long-term variations in climate involved with plate-tectonic changes (Barron 1985). Other scientists have dealt with intermediate-term variations associated with orbital characteristics of the earth's rotation causing glacial and interglacial cycles (Berger 1979; Hays et al. 1976; CLIMAP 1976, 1981). Shorter-term variations of centuries to millennia affecting erosional cycles, vegetational changes, and human history (Wright 1983) also have been reconstructed and identified with volcanic activity cycles, variations in atmospheric gases, and solar activity. Paleoclimatic information emphasizing variations over very short time scales of seasons to centuries can be obtained from the records in ice cores (Dansgaard et al. 1971; Thompson et al. 1984, 1985, 1986), varved sediments (O'Sullivan 1983; Anderson et al. 1985; DeVries 1988; Baumgartner et al. 1989), corals (Lough and Barnes 1990), and tree rings (Bradley 1985; Brubaker and Cook 1983; Fritts 1976; National Academy of Sciences 1975; Schweingruber 1988; Stockton et al. 1985). These high-frequency variations have been identified with ENSO events, yearly solar variations, volcanic dust veils, biennial oscillation of the atmosphere, and simply random climatic variability.

Studies of historical data (Lamb 1977; Ladurie 1971; Lawson 1974; Ludlum 1966, 1968; Catchpole 1985) also provide valuable information on short-term climatic variations. For example, useful climatic information can be extracted by means of content analysis (Catchpole et al. 1970) from diary entries (Baron 1980), ships' logs (Lamb 1977), and a variety of commercial records (Catchpole 1985). These data not only add to the history of climatic variation, but record the influence of climatic variations on human activities (Lawson 1974; Baron 1980). The accuracy and validity of these historical records can be ensured only by a careful interdisciplinary study involving concepts from climatology, biometeorology, communications research, statistics, computer science, and history (Baron 1980). Since few of these interdisciplinary studies have been funded adequately in the United States, a detailed historical account of past climatic variations in North America remains an unrealized potential.

C. Short-term Climatic Variability

Instrumental climatic measurements, historical documents, and paleoclimatic data on short-term climatic variations vary in quality, geographic coverage, and time resolution, as well as length of record. Instrumental data have the highest quality and resolution of the three data types, but few North American climatic records completely span the last two centuries. Bradley (1976a, 1976b, 1980), Bradley et al. (1987), Diaz (1983, 1986), Mitchell (1976), Jones et al. (1982, 1986), Landsberg (1967), and Landsberg et al. (1963) are only a few of the many workers who have examined the longer climatic records. These data provide very uneven coverage over the North American continent during the nineteenth century. A gridded temperature data set (Jones et al. 1986) had 57 percent of the maximum number of points for the Northern Hemisphere in the 1950s, 30 percent in 1891, and only 8 percent in 1851 (Bradley 1988). The pioneering works of Wahl (1968), Wahl and Lawson (1970), Lawson (1974), and Bradley (1976a, 1976b, 1980, 1988) attempt to analyze large-scale climatic patterns, but the poor geographic distribution of nineteenth-century instrumental records has hindered rigorous analysis of these patterns (Gates and Mintz 1975). The spatial coverage of the global meteorological network increases steadily throughout the twentieth century. Although gaps still exist, particularly in the Southern Hemisphere, sufficient data have become available since 1970 for use in numerical models that simulate the main characteristics of the global patterns (Hecht 1985; Schlesinger and Mitchell 1987; Dickinson 1989).

Validated historical references to climatic variations for North America provide a very fragmented record, especially for the western United States. A limited number of well-dated varved sediments provide information on annual climatic variation (O'Sullivan 1983; Anderson et al. 1985; Baumgartner et al. 1989). Corals may also yield information on yearly climatic variations (Lough and Barnes 1990), but these organisms are restricted to subtropical waters. Layered ice cores may show climatic variations on annual time scales (Thompson et al. 1985, 1986) but are confined to high latitudes or high altitudes and have not been reported for the forty-eight contiguous states of the U.S. My study uses tree-ring data from western North America as substitutes for the seasonal and annual averages of instrumental data (Schulman 1956; Fritts 1976).

Little historical and paleoclimatic information of climatic variations over seasons to centuries has been used in climatic simulation studies, perhaps because there are few continuous time series and little spatial coverage of this information when and where the instrumental record is inadequate. If high-frequency excursions in past climatic conditions have exceeded the twentieth-century variability, they could be missing from the simulations. Yet variations on these time scales can have a large impact on a variety of social and economic institutions, biomass production, and agricultural capacity (Pittock and Nix 1986; Liverman et al. 1986; Parry and Carter 1989). At these time scales the effects of many global processes on human activity are most pronounced (Earth System Science Committee 1988).

Increasing evidence suggests that the climate of the twentieth century has been anomalous from that of the long-term average (National Academy of Sciences 1975; Bryson 1974; Bryson and Hare 1974). An awareness of humankind's possible impact on climate has evolved over the twentieth century (Kellogg 1987), with concern focusing in the past decade on possible global warming from greenhouse gases (Hansen et al. 1988; Schneider 1989). The human impact on climate could be intimately related to the twentieth-century warming (Wood 1988b), and a well-reconstructed record for the seventeenth through nineteenth centuries could provide baseline information to evaluate these changes, to understand the climatic variations to expect in the future, and to relate regional variations to global patterns. Tree-ring patterns are one type of biosphere response to climatic variations (Fritts et al. 1991), and tree-ring reconstructions filter out the climatic influences unimportant to tree growth and retain only the climatic variability associated with growth. Climatic reconstructions were derived from time series of annual tree-ring width variations from western North America to help fill in the gaps in climatic information over the seventeenth through nineteenth centuries. The work began early in the 1970s when a team from the Laboratory of Tree-Ring Research at the University of Arizona began a systematic recollecting of arid-site tree-ring chronologies from the western United States, southwestern Canada, and northern Mexico. A grid of climate-sensitive and high-quality tree-ring chronologies was established (Fritts and Shatz 1975), and these chronologies were used for all analyses. The tree-ring grid provided a continuous paleoclimatic sequence reflecting climatic variations, and each element of the data set was dendrochronologically dated to the exact year in which each ring grew (Stokes and Smiley 1968; Fritts 1976; Brubaker and Cook 1983; Stockton et al. 1985). The widths of the dated rings were measured, entered in a computer file, and standardized for subsequent analysis (Fritts 1976).

Many new chronologies have been developed since that time, and a variety of other grids has been assembled (Brubaker and Cook 1983; Stockton and Meko 1975, 1983; Stockton et al. 1985; Hughes 1987). Some of the newer data sets also contained information on ring-density changes (Parker and Henoch 1971; Cleaveland 1986; Schweingruber 1988), which, along with ring widths, should yield more information on past variations in climate (Briffa, Jones, and Schweingruber 1988, 1991). Additional paleoclimatic information was collected from widely scattered sources and sites (Fritts 1987), but the strategy and techniques for integrating these disparate data sets have not been worked out. To maintain a continuity in the analysis, no tree-ring data that became available after the Fritts and Shatz grid was established were considered in this particular work.

D. Climatic Reconstruction

This study employs statistical techniques to deduce spatial fields of climatic information from spatial fields of ring-width variations using trees with rings that are highly limited by climatic conditions. No well-developed theory or statistical guidelines were available for this type of analysis, so the study included (1) diagnosis of each problem encountered in the analysis, (2) identification of alternative solutions, (3) evaluation and selection of the most viable alternative, (4) continuation to the next diagnosis and solution until the best acceptable reconstruction could be obtained, (5) evaluation of the final reconstructions, and (6) comparison of these reconstructions to other information on high-frequency climatic variation and change. This book describes the most important experiments and analyses that were conducted over approximately fifteen years, documents the procedures used to obtain the final climatic reconstructions, and applies these data to some climatic questions and problems. The temperature and precipitation reconstructions are available from the University of Arizona Press or from the National Geophysical Data Center. They are written on floppy disk and can be accessed with a FORTRAN mapping utility in DOS that averages, differences, and maps the seasonal or annual reconstructions over the United States and southwestern Canada (see App. 3; Fritts and Shao in press).

Until the late 1980s, one climatic or hydrologic record usually was reconstructed from one or more tree-ring chronologies (Schulman 1951, 1956; Fritts 1976; LaMarche 1978; Stockton and Jacoby 1976; Cook and Jacoby 1977, 1979; Garfinkel and Brubaker 1980; Hughes et al. 1982, 1984). Often more than one weather record or tree-ring chronology was averaged to enhance the common information on climate (Michaelsen et al. 1987; Briffa, Jones, et al. 1988). The annual variations in the tree-ring chronologies were calibrated with corresponding variations in the climatic records for a particular period of time called the *calibration* or *dependent period* (Fritts 1976; Stockton et al. 1985; Fritts, Guiot, and Gordon 1990).

The calibration estimated a linear equation that transferred one or more predictor variables of tree-ring chronologies to reconstructions of one or more predictand variables corresponding to climatic records (Lofgren and Hunt 1982; Fritts, Guiot, and Gordon 1990). The calibration produced a table (Stockton and Fritts 1971) or statistical equation that could be applied to the tree-ring data outside of the calibration period (the independent period) to estimate or statistically reconstruct the variations in climate. The amount of similarity between the climatic data and the tree-ring estimate was measured over the dependent (calibration) period and expressed as fractional variance (Fritts 1976; Fritts, Guiot, and Gordon 1990). Climatic data from the independent period that had not been included in the calibration were then compared to the reconstructions and used to validate or verify that the independent reconstructions are unlikely to have occurred from a chance relationship.

Multivariate transfer models can calibrate more than one predictor variable with more than one predictand variable (Hughes et al. 1982). For example, Glahn (1968) and Blasing (1978) describe canonical regression as a multivariate technique that is especially well suited for calibrating two large spatial grids. Later, Briffa, Jones, and Schweingruber (1988) and Schweingruber et al. (1991) used a principal component regression model that produced reconstructions equivalent to the canonical analysis.

The following work uses canonical regression models to estimate three climatic variables over five spatial grids from low-elevation, arid-site tree-ring chronologies throughout the North American West. Many models with different structure were obtained in the analysis, so it was also necessary to diagnose the strengths and weaknesses of these models, select the most reliable reconstructions, find a way to combine the best reconstructions, and assess the reliability of the combinations. Finally, the best reconstructions were mapped, plotted, and analyzed to evaluate their reliability over both time and space.

E. Published Reports on the Reconstructions

A brief review of published papers illustrates the kinds of information in these spatial climatic reconstructions and points out the need for this book. The multivariate transfer function using canonical regression analysis was originally described by Fritts et al. (1971). Blasing (1978) developed the necessary software, and Fritts and Blasing (1974), Blasing and Fritts (1976), Fritts (1977, 1978), and Fritts et al. (1977) described the basic methods and reported some of the earliest results.

Blasing and Fritts (1975) used these transfer-function techniques to reconstruct summer sea-level pressure anomalies along the American Arctic from arid-site tree-ring chronologies in western North America. They then examined independent tree-ring data from forests that border the American Arctic to see whether the large-scale features in the pressure reconstructions could be associated with tree-growth variations along the Arctic tree line. Only tentative inferences could be made because the reconstruction work was not complete, and other types of proxy data on climatic variation in the region had not been examined and tested.

Fritts and Lofgren (1978), responding to questions from the United States Army Coastal Engineering Research Center, reported that large synoptic-scale variations in climate were reconstructed during the seventeenth through nineteenth centuries which do not necessarily coincide with ideas about worldwide average changes in climate. These data suggested that realistic planning for the future might better center on seasonal climate and the larger variance of information in the smaller regions, rather than focusing exclusively on worldwide changes varying only on time scales of centuries to millennia.

Fritts, De Witt, et al. (1979) corrected annual precipitation statistics, using estimated long-term statistics from dendroclimatic reconstructions for six regions of the United States, to assist the assessment of government plans for nuclear waste management. They found average precipitation before the twentieth century to have been below the twentieth-century figures for three regions in the American Southwest and above the twentieth-century figures for one region in the Northwest and two regions in the East, and the variability of precipitation was reconstructed to have been higher in the past three centuries than in the present.

Fritts (1984) reported that the first sixty years of the twentieth century were slightly cooler and more moist in the West than the reconstructions from the seventeenth through nineteenth centuries but warmer and dryer for the central and eastern regions. A comparison of these data with the sea-level pressure reconstructions suggested that more storms traveled farther south off the California coast beginning around the turn of the century and moved, on the average, in a northeast direction through the Great Basin to the Great Lakes, bringing cooler and more moist conditions to the West with more southerly flow and warmth to the Midwest. Projections for the future that were based solely on the twentieth-century records would overestimate the values for moisture in the Southwest, underestimate temperature west of the Rocky Mountains, and overestimate temperature to the east. Lower variability occurred in twentieth-century precipitation. Reconstructions of this kind should be used to extend the baseline information on past climatic variations so that projections for the future include a more realistic estimate of natural climatic variability than is available from the short instrumental record.

Fritts and Lough (1985) examined the most reliable large-scale features and low-frequency variations of temperature reconstructions by averaging the data over space, filtering the data over time, and comparing these series to the averages of instrumental records of Jones et al. (1982), Manley (1974), and Groveman and Landsberg (1979), as well as to upper tree-line and Arctic dendrochronologies. Although the twentieth-century warming is a prominent feature of all reconstructed temperature plots, in some earlier time periods the filtered temperature estimates were as high as or higher than temperatures in the twentieth century, particularly in western North America (Boden et al. 1990).

Fritts and Lough (1985) also noted that the correlation with the Jones et al. (1982) hemispheric temperature record was strong from 1881 to 1946 and was weak after 1946, when hemispheric temperatures temporarily declined. The Fritts and Lough series was inversely correlated with Manley's temperature series around the middle of the eighteenth century but not for any other period, which suggested that the association of long-term temperature trends between these two regions was no greater than that expected from chance variation. The correlations with the Groveman and Landsberg (1979) temperature reconstructions were positive and significant only during 1656 to 1680 and 1915 to 1940. The greatest disagreement occurred when Groveman and Landsberg's proxy data came largely from the European and North Atlantic sectors of the hemisphere and not from the United States. The comparison suggested that the Groveman and Landsberg reconstructions could be made more reliable by using high-quality, continuous, and well-dated tree-ring series and their derived reconstructions averaged over the United States. Fritts and Lough (1985) proposed that a variety of proxy information on large-scale climatic variations should be compiled and compared to evaluate both regional and global average changes. Then all of the evidence should be examined from a climatological viewpoint to evaluate the most important similarities between various climatic sensors and to infer where and when past climatic variations actually occurred.

Lough et al. (1987) note a statistically significant association between drought or flood in China, precipitation in North America, and large-scale sea-level pressure variations over the North Pacific only when Chinese precipitation was lagged one year behind the North Pacific sea-level pressure and North American precipitation. The same relationships were also found between the early Chinese documentary records and dendroclimatic reconstructions of both annual sea-level pressure and precipitation, and the patterns identified in the twentieth-century were also present during the earlier periods.

Lough and Fritts (1985) used the climatic reconstructions to study ENSO relationships and then calibrated tree-ring data with the Southern Oscillation (SO). Significant high-frequency spectral peaks were noted at periods from 2 to 10 years like those found in the SO; low-index events were reported to have been less frequent during the nineteenth century than before or since; and extreme SO events reconstructed in 1792–1793 and 1815–1816 may have been of comparable magnitude to the extreme 1982–1983 event.

Lough and Fritts (1987) also examined the possible impact of past explosive volcanic eruptions on climate by relating the eruption events to the tree-ring-derived reconstructions. It was found that the average annual temperature declined in the central and eastern United States following low-latitude eruptions. Winters and summers were warmer in the western United States following low-latitude eruptions, however, and cooling was most marked in spring and summer throughout the eastern and central parts of the country.

F. The Analysis Strategy

Choices of the appropriate tools and of the most viable reconstructions were based on various statistics obtained from the procedures of calibration, verification, reconstruction, model selection, and merging.

The sheer size of the task before us required the following strategies:

(1) The work proceeded by asking one question at a time, experimenting by varying one factor at a time, selecting the most promising outcome of the experiment, and then proceeding to the next question.

(2) As more and more questions were resolved and new developments were reported in the literature, it was a temptation to reopen and reevaluate earlier questions. Recognizing that it was important to complete and assess the results from at least one full round in the analysis, we resisted this temptation and postponed the reanalysis until the first round was completed and reported. This book is that report. For this reason some alternative strategies that were not considered are listed in the recommendations to aid future investigations.

(3) Likewise it was decided that the same tree-ring and climatic data sets should be used throughout the analysis so it would be possible to compare the statistics at each stage in the analysis to evaluate technique improvements. This approach allowed a thorough diagnosis of the problem over a large data set. It did not allow deleting of data points once the grid was established. Thus, the average statistics from the grid did not yield the highest average statistics that were possible, because the data points that were poorly fit could not be excluded from the average statistics at a later date. The spatial variation of the statistics and differences in model structure were preserved to evaluate the model capabilities and the spatial extent of valid reconstructions. It is hoped that the discriminating reader will keep this in mind when comparing these averaged results to other dendrochronological studies aimed only at maximizing a particular relationship.

(4) The maximum distance over which significant growth-climate relationships might be expressed was unknown when the project began. Two grid sizes for temperature and precipitation were used in the analysis. The smaller grids included climatic stations in western North America covering approximately the same area as the tree-ring grid. The larger grids included eastern United States stations as well as the stations from the West. These grids were subdivided into climatic regions and the differences diagnosed to evaluate the important relationships involving space. These results were also used to identify some strengths and weaknesses of the models that were tested.

2

Characteristics of
the Tree-Ring Data

The primary goal of collecting tree rings for dendroclimatic analysis is to obtain the longest and clearest record of past climatic variations (Schulman 1956). Ring-width measurements do not always contain information on climate, however. It is necessary to sample sites on which climatic conditions are most likely to have limited processes affecting growth (Fritts 1976; LaMarche 1982). In addition, dendrochronologists often search for the oldest trees growing in the area to obtain the longest possible record (Fritts 1982). Fortunately, the oldest trees on such sites often contain the highest-quality records.

Not all climatic factors are directly limiting to the growth-controlling process. Nonlimiting factors would not be recorded in ring widths unless they are correlated with other climatic conditions that are growth limiting (Fritts 1976). This chapter describes the characteristics of tree-ring data and some of the biological relationships governing the ring response to variations in climate.

A. Chronology Development

The ring-width data were sampled from semiarid-site trees of *Pseudotsuga menziesii*, *P. macrocarpa*, *Pinus ponderosa*, *P. edulis*, *P. flexilis*, *P. longaeva*, *P. jeffreyi*, and *Abies concolor* (Fritts and Shatz 1975; Fritts 1976). The most stressful sites were located, the most mature and potentially responsive trees were identified, and the ring characteristics sampled by extracting a core from the tree bole (Schulman 1956; Stokes and Smiley 1968; Fritts 1976). If the ring widths were variable, the growth trend associated with increasing tree age was normal, no evidence appeared of disturbance or injury, and no scar or pitch pocket interrupted the continuity of the ring-width time series, then the tree was recorded and a second core extracted from the opposite side of the stem (Fritts 1976). Analysis of variance indicated that a sample of paired cores from ten to twenty-five trees was adequate (Fritts 1976). At the laboratory the cores were dried, mounted, and surfaced with sandpaper to facilitate visual examination. Each ring was dated as to the exact year in which it was formed (Fritts 1976, 1986; Stokes and Smiley 1968; Dean 1978), and the widths were measured.

Ring-width measurements are nonstationary time series since they exhibit declining means and variances associated with increasing tree age (Fritts 1976; Monserud 1986). For this reason, width measurements must be transformed into standardized ring-width chronologies that have stationary means and homogeneous variance. A negative exponential function or straight line with a negative slope is fit to each ring-width series (Fritts et al. 1969; Graybill 1979a, 1979b; Cook 1987), and the ring-width measurement for each year is divided by the curve estimate to obtain a ring-width index. The indices for each year are averaged for all trees of a given species from a single site to obtain a site chronology (Fritts 1976; Fritts and Swetnam 1989; Brubaker and Cook 1983).

Standardization removes large amounts of nonclimatic variations unique to individual trees and sites and preserves a large part of the climatic variations in ring-width measurements that are common to trees in a particular region. The nonclimatic variations in ring widths are associated with the geometry of the ring, increasing age of the trees, individual tree histories, local stand conditions, and site factors. Standardization could remove some very long-term climatic trends resembling the curves fit to the ring-width measurements. This problem was minimized by fitting a relatively rigid exponential function with one inflection so that the wavelength of the removed variation generally exceeded the age of the sampled trees. If we assume that the average tree age is approximately two hundred years, the average time scale of the removed variation would be expected to have been greater than that value. Power spectra (Blackman and Tukey 1958) of the chronology data confirmed that variance, for time scales of 2 to 200 years, was well preserved in these standardization tree-ring chronologies.

B. Chronology Selection

There were approximately six hundred candidate ring-width chronologies available throughout the North American West in 1969, when this study was initiated. The best chronologies were selected from an area of 4,620,000 km^2 covering thirteen western U.S. states, three Mexican states, and two Canadian provinces (Fritts and Shatz 1975). The candidate chronologies were grouped by area and ranked according to statistics that indicated their potential for climatic reconstruction work. Those with the highest standard deviations, highest mean sensitivities, highest percentage of variances in common, and lowest first-order autocorrelations received the highest rankings (Fritts and Shatz 1975; Fritts 1976).

The variance at different frequencies was examined by applying high-pass and low-pass digital filters to the chronologies (Fritts 1976). Marked differences in standard deviations and unusual differences in the correlations between the high-pass and low-pass filtered components in chronologies from neighboring sites were used to identify those time series with substantial nonclimatic information. This procedure assumed that the low-frequency variations found in only one or two chronologies from an area were the results of site disturbance or other local factors rather than regional climatic patterns. Chronologies with these unusual time series characteristics were eliminated so that the selected chronologies had more or less equal amounts of high- and low-frequency variance in common.

When a number of candidate chronologies were available from the same locality and the chronologies were highly correlated with one another, usually one was selected on the basis of its rank and length. This method reduced some of the redundancy in the grid but did not even out the density completely, because in some regions there were too few candidate chronologies to make an even grid possible without reducing the overall density to an unacceptably low level. We assumed that it was better to use a variety of chronologies wherever possible to maximize the information on climate at certain grid points than to reduce the climatic information to a level dictated by the area of least chronology coverage.

C. Basic Statistics

A grid of 89 chronologies was selected that had continuous time series for 1700 to 1963, an average density of 7.8 sites/Mkm^2 (Cropper and Fritts 1982), and a relatively even distribution of chronologies over the North American West. An additional 13 chronologies that were included by Fritts and Shatz (1975) to provide a denser set of chronologies in and around Arizona were not considered. However, there was a 65-chronology subset of the 89-chronology grid that had a continuous

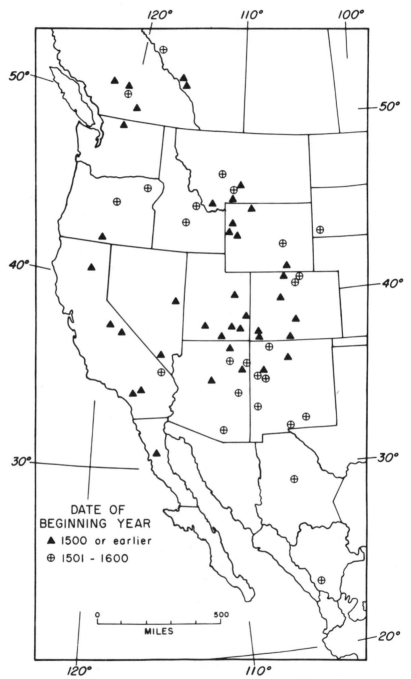

Fig. 2.1. Locations of the 65 chronology sites used in this work (Fritts and Shatz 1975).

Table 2.1. Numbers of Chronologies Selected from Each Species for the Three Study Grids

| Species | Grid Designations and Interval in Common | | |
| | 40 | 65 | 89 |
	1500–1963	1600–1963	1700–1963
Abies concolor	0	1	1
Pinus edulis	4	6	11
Pinus flexilis	5	5	5
Pinus jeffreyi	2	2	2
Pinus longaeva	3	3	3
Pinus monophylla	0	0	1
Pinus ponderosa	6	16	21
Pseudotsuga macrocarpa	1	1	1
Pseudotsuga menziesii	19	31	44
Total	40	65	89

Source: Fritts and Shatz (1975).

time series for 1600 to 1963 and had an average density of 7.1 sites/Mkm^2. An additional subset of 40 chronologies spanned 1500 to 1963 and had an average density of 4.7 sites/Mkm^2. The different symbols in Figure 2.1 show the distribution of the chronologies in the second and third subsets. Only one species was used in an individual chronology, but chronologies from nine different species were included in the 89-chronology grid (Table 2.1).

Stockton and Meko (1975, 1983) developed a 40-chronology network for reconstructing Palmer Drought Severity Indices (PDSI) in the central and western United States which represented the same grid as the 89-site grid. The authors basically eliminated 49 chronologies to obtain an even distribution; unlike the case in 40-chronology subset described above, Stockton and Meko's chronologies were continuous only from 1700 to 1963. Stockton (personal communication 1987) acknowledged later that this decision to emphasize an even distribution of sites not only produced a shorter record, but yielded inferior reconstructions of climatic variation. That is why we wished to evaluate grid densities before embarking on the full analysis.

Briffa, Jones, and Schweingruber (1988) used a network of maximum latewood density chronologies from 37 European sites to reconstruct

Table 2.2. Average Percentage of Variance Calibrated in Sea-level Pressure Using the Three Chronology Grids

Number of Chronologies	Season				
	Winter	Spring	Summer	Autumn	Average
89	28.7	38.1	*38.9*	30.1	34.0
65	*36.0*	*40.1*	38.6	*36.4*	*37.8*
40	21.9	20.6	25.2	25.7	23.3
Average	28.9	32.9	34.2	30.7	31.7

Note: Four couplets used for autumn, three used for other seasons. Maximum values are in italics.

Table 2.3. Average Percentage of Variance Calibrated in Temperature and Precipitation Using Twelve Couplet Models of Different Structure Applied to the Three Chronology Grids

Variable	Number of Chronologies	Season				
		Winter	Spring	Summer	Autumn	Average
Temperature[a]	89	21.1	35.2	38.6	*33.7*	32.2
	65	*30.8*	34.3	*39.2*	27.9	*33.1*
	40	25.0	*35.7*	31.7	31.1	30.9
	Average	25.6	35.1	36.5	30.9	32.0
Precipitation[b]	89	39.0	37.5	25.7	*31.8*	33.5
	65	*41.0*	*37.6*	*27.3*	30.6	*34.1*
	40	34.0	38.6	26.3	27.3	31.6
	Average	38.0	37.9	26.4	29.9	33.1
	Grand Average	31.8	36.5	31.4	30.4	32.6

Note: Maximum values are in italics.
[a]Using 15 eigenvectors of the 46-station temperature grid.
[b]Using 20 eigenvectors of the 52-station precipitation grid.

summer temperatures over a meteorological grid. Their network covered a comparable geographic area, was continuous from 1750 to 1975, and had a density of 3.6 sites/Mkm^2, which was half the density of the 65-chronology subgroup. Schweingruber (1988) described a new network for western North America. Other tree-ring networks have been described as points on a map (Brubaker and Cook 1983; Stockton et al. 1985; Hughes 1987) but have not been thoroughly evaluated, combined into a network, or used to reconstruct climate.

A network of climatic stations, described by Kutzbach and Guetter (1980), was designed to simulate the calibration, verification, and reconstruction of climate from a widely distributed tree-ring grid. Kutzbach and Guetter wished to determine guidelines regarding spatial density and locations of the data points necessary to maximize the information in climatic reconstructions of continental- to hemispheric-scale variations in climate. The authors used different size and density grids of temperature and precipitation as if they were tree-ring data to reconstruct different size and density grids of sea-level pressure. The highest-density grid was 2.5 sites/Mkm^2, which was approximately half the density of the 40-chronology grid and one-third the densities of the 65- and 89-chronology grids. The densities of the 65- and the 89-grids of Fritts and Shatz (1975) were about three times the densities used in other dendroclimatic work. Since these grids of chronologies were denser than those used for any other dendroclimatic reconstruction, care was taken to evaluate and select methodologies that could extract a diverse amount of climatic information from such a varied and dense dendroclimatic data set.

It was hoped at first that a full analysis would be made on each of the three chronology subsets so that the effects of chronology density could be tested. As the complexity of the task grew, it was necessary to reduce the scope of this investigation to only one subset. Each of three subsets was evaluated by using three or four couplet calibration models (see Chaps. 4 and 5) of different structure to relate seasonal sea-level pressure to the tree-ring chronology variations. The average calibrated variance was tabulated (Table 2.2), but the program to calculate verification statistics had not been developed. The same analyses were performed on temperature and precipitation, but twelve different couplet models were used for these analyses (Table 2.3; see Chaps. 4 and 5 for more discussion of couplets and choices related to model structure).

The average calibrated variance for all four seasons (Tables 2.2 and 2.3, last column on the right) was always highest for the 65 chronologies, and it was highest for the 65 chronologies for eight out of twelve seasons. The 89 chronologies had the next highest calibrated variance, and the 40 chronologies had the lowest. The 65 chronologies not only provided the most climatic information of the three data sets, but they were 100 years longer than the 89 chronologies. Although the 40 chronologies would have provided a longer record, it seemed unwise to use them because the lower calibration might increase the error and limit the diagnosis and analysis of dendroclimatic procedures. The 65-chronology grid (Fig. 2.1) was selected for all subsequent calculations.

A common omission in many dendroclimatic publications is a lack of information on the characteristics of tree-ring data used in analyses (Stockton et al. 1985). It is easy to list species, latitudes, longitudes, and record lengths and to plot points on a map, but it is more difficult to document the quality of the tree-ring data, including specimen depth, dating, and some basic statistics that indicate the quality of the data set. The following description details the statistical qualities of the ring-width data used in the 65-chronology set.

The standard deviation estimates the variability in a chronology without regard to its frequency. It is a gross measurement of variance that provides potential information about climatic variation. The average standard deviation for the chronologies is 0.37 with ± 0.18 95 percent limits (Fritts and Shatz 1975). Autocorrelation estimates the variability that is correlated between lags in the time series, which reflects low-frequency variations, including trend. The first-order autocorrelations for the chronologies average 0.42 with ± 0.21 95 percent limits (Fritts and Shatz 1975). This finding indicates that substantial but not excessive variance is present for time periods of 2 or more years. Mean sensitivity measures the relative variance at high frequencies or periods less than 2 years in length. The average mean sensitivity is 0.35 with ± 0.22 95 percent limits. These three statistics indicate a relatively high amount of year-to-year variation in these arid-state tree-ring chronologies and suggest that they should contain considerable high-frequency information about past climatic variation.

Other statistics reveal different features of the data set. Tests of normality show that approximately 90 percent of the chronologies are normally distributed at the 95 percent confidence level. Thus most, but not all, of the chronologies are well behaved and can be assumed to be normally distributed. The general form of the variance spectra (Blackman and Tukey 1958) for the 65 chronologies resembled red-noise spectra that are typically found in physical systems driven by climatic variations (Bryson 1974; Bryson and Dutton 1961; Mitchell 1976). These tree-ring chronologies do appear to integrate the climate for 1 or more years, and that is probably an important contributor to the red-noise spectra of tree-ring chronologies. Not all of the red noise can be attributed to climatic variation, however. In spite of the careful screening of these chronologies, some exogenous and endogenous growth disturbances are undoubtedly present, and these effects are probably important contributors to the red-noise component (Graybill 1982; Cook 1987; Fritts and Swetnam 1989).

D. Variance in Common, the Error, and the *SN* Ratio

Analysis of variance or correlation analysis can estimate the variance common to a set of chronologies (Fritts 1976; Wigley et al. 1984). This common variance may be attributed primarily to limiting climatic conditions, and it was estimated by DeWitt and Ames (1978) to be approximately 60 percent for arid-site trees in western North America. Most ring-width chronologies from Arctic regions, central Europe, and the eastern United States have substantially less variance in common than these low-elevation, arid-site chronologies. This statistic is not commonly reported for densitometric chronologies, but the excellent correlation between chronologies in different areas (Schweingruber et al. 1979; Briffa, Jones, and Schweingruber 1988) suggests that the percentage of variance in common for densitometric data for nonarid areas may be more comparable to the variance in the ring-width chronologies of the North American West than to the variance in ring-width chronologies from less arid regions.

Pittock (1982) states that ideally climatologists would like to reconstruct at least 70 to 80 percent of the instrumental record variance, and he complains that often only a quarter or a third of the variance that is common to all cores and trees can be statistically related to climatic variation. He contends that this shortfall of explained climatic variance is due to (1) nonclimatic factors present in the common tree-ring variance, (2) insufficient replication of sampling so that there are large errors in the chronology estimates, (3) too great a distance between the climate data and the tree sites, (4) omission of important climatic variables from the statistical analysis, and (5) inadequate model structure, including nonlinear effects.

Other concerns should also be on his list. The so-called shortfall may also be attributable to (6) errors in the climatic data, (7) differences between the microsites of the climatic stations and the tree sites (the former are frequently in valleys and the latter in the mountains), (8) integration of the climatic record by the tree's responses, which smooth out and virtually eliminate a large amount of high-frequency variance, and (9) climate factors important to the tree that are not measured adequately by existing meteorological instruments (Fritts 1976). Thus, there are both climatic and dendrochronological reasons for the low correlation between climatic variation and tree-ring data.

In addition, Cropper (1982) points out that Pittock (1982) made his comparisons by using the variance in the individual core measurements rather than the error of the mean chronology. Cropper (1982) and Fritts (1976) showed that the error in a chronology is dependent on the variance within and between trees. If we disregard the error within trees for the moment, we can obtain the chronology error by dividing the variance attributed to differences between trees (D), that is, the total variance minus the variance in common, by the number of trees in the sample (N). The variance in common among trees (C), which is analogous to the signal in an electronic system, can be estimated by analysis of variance (Fritts 1976); C can then be divided by D and multiplied by N to obtain the signal-to-noise (SN) ratio.

$$SN = CN/D \qquad (2.1)$$

This equation is an oversimplification because it ignores the effects of multiple cores from the same trees. It is sufficient, however, to illustrate the effects of varying the numbers of trees on the error and the potential climatic signal in the 65 tree-ring chronologies.

The signal or variance that is common to all trees (Fritts 1976) is unbiased by the sample size and is unaffected by the number of trees on which the chronology is based, while the noise varies inversely with the sample size, so that the SN ratio expresses the relative information on climatic variation in a chronology. A definitive work by Wigley et al. (1984) develops these concepts further and describes equations that use the correlation between core chronologies in different trees to estimate the relative signal strength in a subsample. This work was not available when these 65 chronologies were evaluated by Fritts and Shatz (1975), so the percentage of the analysis of variance estimates for the variance not in common (differences) between trees as opposed to the percentage of variance in common between trees (Fritts 1976) is used with the number of cores to estimate the signal strength in this example.

Using the relationship in Equation 2.1 and assuming that only one core was collected from each tree, we can approximate the relative error over the grid, ϵ_t, for each year t as:

$$\epsilon = \frac{1}{65} \sum_{65}^{1} v_t / \sqrt{N} \qquad (2.2)$$

where v_t is the square root of the variance of the differences between trees (variance not in common) for the site chronology.

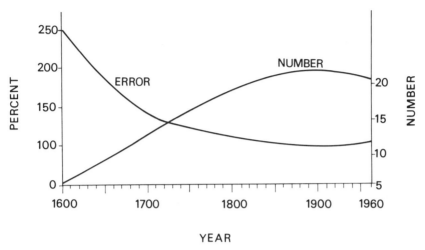

Fig. 2.2. The average number of cores and the estimated average error of the 65 standardized chronologies plotted as a function of the year in which the rings were formed.

The relative error over the entire grid (Fig. 2.2) in 1600 was as much as 250 percent of the error for 1901–1963, the base period used for comparison. As the sample size increases, the error decreases to 200 percent by 1640 and 150 percent in 1690. The relative error decreases slowly until 1800, after which little change in the error is evident. Thus, the errors in the earliest portions of the chronologies from the seventeenth century are one and a half to two and a half times larger than the errors in the twentieth century, the interval used in the calibration. Because the early portions of tree-ring chronologies are usually based on fewer trees, errors will be higher and any derived climatic reconstructions will have less reliability.

Nonetheless, the relative amount of error in the 1600s is probably overestimated and the signal underestimated in these calculations, because at the beginning of a chronology a greater proportion of the measurements comes from different trees than from opposite sides of the same trees compared to samples from the latter part of the chronology. The 65 chronologies were based on an average of 6 cores in 1600, 13 cores in 1700, 19 cores in 1800, and 22 cores in 1900 (Fig. 2.2). Although most of the cores in 1600 probably came from different trees, the 13 cores for 1700 may represent 7 to 8 trees on the average with replications on only 5 to 6 individuals, and the 19 cores in 1800 probably came from 10 or 11 trees, on the average, with 8 or 9 replications. This overrepresentation of information from different trees in the early part of the chronology would reduce the slopes of the curves in Figure 2.2 slightly. It would facilitate future analyses of the errors in dendroclimatic grids if the statistics described by Wigley et al. (1984) could be listed along with the average numbers of trees and cores used in each year of the chronologies.

E. The Signal Strength

In addition to estimating the errors, it is useful to evaluate the strength of the signal in the site chronologies. The following analysis assumes that the correlation coefficient is an index of the signal strength. However, another important contributor to the correlation between different chronologies is the relative distance between the tree sites. Increasing the distance between sites would be expected to increase the differences in microclimates of the respective sites, cause more differences in the growth-controlling processes, and reduce the correlation between the chronologies. Therefore, distance as well as signal strength was considered in the following analysis.

Each chronology was paired with every other chronology; the correlations between the values from 1700 through 1963 were calculated, and the values of the correlation coefficients were plotted against the separation distances between the site localities. There was a highly significant inverse relationship between the correlation coefficient and the square root of the distance explaining 72 percent of the variance (Cropper and Fritts 1982). The Y intercept of the regression line corresponded to a correlation value of 0.762 for a separation of zero distance. The coefficient of determination (r^2) was 0.581, indicating that the common signal for chronologies from neighboring sites averaged 58 percent.

The correlations between the chronologies were, on the average, positive and significant at the 95 percent level for distances up to 1,154 km. This relationship suggested that the 65-chronology tree-ring grid could contain information on climate as much as 1,154 km from the boundary of the grid, which would be as far as the Mississippi River. In addition, the correlation coefficients were sometimes significantly negatively correlated at distances of 2,000 km or greater. This finding suggested that anomalously high values of a climatic anomaly were often associated with anomalously low values of the anomaly at distances of 2,000 km, that the half wavelength of the climatic phenomena affecting ring-width growth in western North America may be around 2,000 km, and that information about global climatic variations might be obtained at distances greater than 2,000 km beyond the area covered by the tree-ring chronologies. The enlarged area would cover most of North America, including Alaska and the Canadian Arctic.

The correlation coefficients were highest for a given separation distance when the sites were at similar latitudes, a relationship that may be associated with the movement of storms from west to east across the region. Likewise, the correlation coefficients were lowest for a given separation distance when the sites were at similar longitudes. Additional variation was noted in the correlation of chronologies with increasing separation distance, but no sources of variation were identified in this particular analysis. Some of the possibilities included the geographic proximity of mountain masses and other features, which would create different climatic patterns affecting the growth variations (see Chap. 3, Sec. C); differences in the responses of the trees to the same climatic factors; differences in the time-series characteristics of the chronologies which would influence the covariance spectrum; and differences associated with local stand conditions, disturbances, and the kinds of trees and species that were sampled. Though it is dangerous to make conclusions about distance relationships greater than the diameter of the tree-ring grid, this correlation, as well as other evidence (see Kutzbach and Guetter 1980), supported the possibility that climatic variations may be reconstructed beyond the area of the tree-ring grid. The climatic stations were chosen from a wide geographic area to test this possibility.

F. Constraints to Modeling the Ring-width Response

Many biological cause-and-effect relationships between climatic factors and tree growth are known to exist (Kozlowski 1971; Kramer and Kozlowski 1979), and there is abundant evidence that ring width is often limited by and correlated with variations in climate (Fritts 1976; Schweingruber 1988). This brief review summarizes the most important relationships linking spatial patterns of tree rings to spatial patterns in climate.

A tree-ring width can be regarded as one kind of growth response to climatic conditions operating previously to and concurrently with a particular season of growth. The thickness of a ring varies as a function of (1) the length of the growing season, (2) the rate of cell division, and (3) the amount of cell enlargement (Fritts 1976). All three processes are influenced by biochemical and physical conditions in and around the tissues of the tree. These processes are, in turn, influenced by limiting microenvironmental conditions related to (1) the balance between the water uptake in the roots and water loss from the leaves; (2) the amount of light, temperature, and moisture affecting photosynthesis in the green tissues of the tree; (3) factors controlling the energy balance and respiration processes; and (4) growth-regulating substances (Fritts 1976; Fritts et al. 1991). In addition, internal conditions and processes result from antecedent environmental conditions that have occurred in days, months, and seasons preceding growth. All of these conditions are in turn related to antecedent conditions of measured macroclimate, including temperature, precipitation, solar radiation, wind, humidity, and the amounts of various atmospheric gases.

The degree of linkage and the resulting correlations between the macroclimatic factors and the growth-controlling processes depend on (1) the length of the cause-and-effect chain of events that is involved; (2) how tightly or loosely the surrounding microenvironments are coupled to the processes along the interacting surfaces of the leaves, stems, and roots; (3) how tightly the microenvironment at the plant surface is coupled with the macroclimate of the forest stand; and (4) how tightly the macroclimate of the stand is coupled with the instrumental measurements used to calibrate the growth-climate relationships.

The tightness and structure of the coupling can change as the plant and its competitors grow and develop, as limiting conditions vary from one season to the next, and as human activities influence the climate at the weather-recording station or modify the instruments used to monitor climatic variations. It is assumed that dendrochronological procedures and techniques help to minimize some of the changes in coupling associated with increasing age of the trees, but any decrease or increase in coupling will affect a decrease or increase in the correlated relationship of ring width with climate.

Because there can be many relationships and interactions involved (including genetic, topographic, and geographic differences among sampled species and sites), each ring-width chronology will include (1) some similarities in response found in neighboring chronologies derived from similar species and microhabitats which in part reflect macroclimatic variations, (2) certain differences in response leading to unique variations in all trees sampled from a given site and species, and (3) differences between trees in a given site, chance variations, and errors in ring-width measurement.

The differences in (2) can result from varying responses of the trees to the same macroclimatic variables from one season to the next or from similar responses to microclimatic differences from one chronology site to the next. Theoretically, at least some of this information should be available and could be used, if both the differences and similarities among the different chronologies are considered in the calibration analysis. However, it is difficult to distinguish between differences that represent error in measurement or noise in the relationships and fundamental differences in response.

This problem can be dealt with in at least two ways. The first is to follow Graumlich (1985) and analyze each chronology exhaustively; remove persistence, extreme values, non-normality, and other undesirable characteristics that represent the noise component; and model the individual chronology with individual records of climate. The second is to screen individual chronologies less critically and average chronologies into some large-scale patterns (the PCs) to enhance the variance that is common to more than one chronology. This method reduces the variance that is not common between individual chronologies and that is assumed to be error and nonclimatic effects. Because of the Central Limit Theorem, the combinations are more likely to be normally distributed than the individual series and less likely to be distorted by an extreme chronology index. The second method can reduce a large number of chronologies to a small number of combinations that can be more readily related to patterns of climatic variation. Techniques and procedures must be developed to identify and remove the smaller-scale elements of the model that contribute to insignificant variance. In practice, both approaches should be used. Practicality, scale of the analysis, and the availability of statistical tools and other resources govern the choice.

G. Response-Function Analysis

The results from response-function analysis can be used to identify which climatic factors are correlated with ring-width chronologies. Response-function analysis uses principal component regression (Fritts et al. 1971) to predict statistically the chronology values from monthly temperature and precipitation measurements. Partial regression coefficients are produced that express the relative importance of each monthly climatic variable to ring-width index (Fritts 1976). The work of Draper et al. (1971) and Rencher and Pun (1980) indicates that a bias is present in the probability of the stepwise regression estimates and that it may be because of the stepwise selection process. The technique may underestimate the size of the confidence limits by 66 percent (Cropper 1985), and there is a variety of opinions as to which alternative is best (Hughes et al. 1982; Guiot et al. 1982). Fritts and Wu (1986) compared some alternatives and found that they produced very similar results. In spite of its weaknesses, response-function analysis provides a means of screening a large number of climatic variables such as monthly temper-

Table 2.4. The Percentage of Response Functions with Significant Positive or Negative Coefficients for at Least One Month within a Particular Season

	Coefficient			
	Temperature		Precipitation	
Season	% Positive	% Negative	% Positive	% Negative
Prior summer	35	57	49	40
Autumn	31	69	87	28
Winter	33	57	90	18
Spring	22	73	93	8
June – July	36	71	46	31
Weighted average	31	65	75	25

Note: $p = 0.66$. Responses with significant positive and negative coefficients for different months in the same season are tallied in both columns (Fritts 1974).

ature and precipitation and deciding which variables may be combined or studied more carefully for nonlinear effects (Graumlich and Brubaker 1987). The evidence from 127 response-function analyses (Table 2.4) (Fritts et al. 1971; Fritts 1974, 1976) indicates that the ring width may be correlated with temperature or precipitation in any season and that the relationships can be direct or inverse. The variability between response-function results suggests that different climatic conditions could become limiting to growth in different species and sites and that the effects of a climatic variable can vary from one season to the next.

Monthly precipitation in autumn, winter, and spring preceding growth is coupled most strongly with ring-width index, and the relationship is direct (Table 2.4). Precipitation in the prior and current growing seasons is not as strongly coupled to ring width, whereas monthly temperature during all seasons, including the prior summer months, is moderately coupled and inversely related to ring width.

Approximately 75 percent of the significant response-function coefficients for precipitation are positive, and 25 percent are negative. Only 31 percent of the significant coefficients for temperature are positive, but 65 percent are negative. It is clear from this analysis that growth in semiarid sites is more tightly coupled with local conditions of precipitation than with local conditions of temperature, and that the climate during the prior summer, autumn, winter, and spring is as important to ring-width variations as the climate during late spring and summer concurrent with growth. The relationship with precipitation on these sites is direct, and the relationship with temperature is primarily but not exclusively inverse.

Fritts (1976) reported that the errors in a tree-ring chronology often increase as a function of the yearly index. The implication is that years of low growth have less error than years of high growth, and this difference in error can be transferred to any climatic estimate. Since ring widths in the 65 tree-ring chronologies are directly related to precipitation and inversely related to temperature, low growth would be transferred into estimates of low precipitation and high temperatures, and estimates of dry and warm conditions versus wet and cool conditions would be more precise. Because of less error, estimates of drought or extreme warmth are more likely to pass significance tests than estimates of wet conditions or a cool climate.

Although random variations and error contribute to large chronology variance and wide confidence bands in the response-function results, some differences in response appear to be associated with differences in species, exposure, altitude, and geographic location of the sites (Fritts 1974, 1976; Brubaker 1980). A flexible multivariate model that could relate differences in tree-growth response over space to seasonal rather than annual variations in climate seemed appropriate for this research (Fritts et al. 1971).

Also contributing to differences in response are important effects of climate that have occurred in previous years but require time before they influence growth, and preconditioning phenomena that may or may not be correlated with prior growth and that influence the capacity of the trees to respond to climate. Both effects contribute to the red-noise spectrum that has been noted in chronology results (Bryson and Dutton 1961; Fritts 1976; Monserud 1986; Rose 1983). Possible modeling structures to handle these features include (1) lagging ring width one or more years behind the climate, (2) including the climate of the prior year in the modeled relationship, and (3) removing first-order autocorrelation from each chronology (see Fritts 1976 for more discussion on lags and leads in response-function analysis).

H. Correcting for Autocorrelation

The developing theory of time-series analyses, first described by Jenkins and Watts (1968), offers some alternatives for dealing with autocorrelated tree-ring chronologies (Meko 1981; Cook 1985; Graumlich 1985). These techniques become highly complex when they are applied to spatial analysis (Bennett 1979). The alternative that was selected at the beginning of this study was simply to remove autocorrelation as follows:

$$C_t = c_t - \tau(c_{t-1} - \overline{m}) \qquad (2.3)$$

where τ is the autocorrelation at lag one, C_t is the chronology index for year t corrected for first-order autocorrelation, c_t is the uncorrected chronology index for year t, and $\overline{m}$ is the mean of all indices. Two tree-ring data representations of the 65-chronology grid were used in the modeling. The first-order autocorrelation was removed from one, and the other was untreated.

3

The Climatic Data

An important decision early in this study was to identify climatic data sets that would provide a reasonable spatial coverage of the area centered over the 65 western North American chronologies. The continuity of the climatic data over time was also important, along with the overlap with the tree-ring chronologies, which should be as long as possible to obtain stable transfer-function results. Many good temperature and precipitation records from the United States and Canada began in 1901, with record fragments extending back into the nineteenth century. In addition, a grid of monthly sea-level pressure starting in 1900 was available on magnetic tape. Also the three data sets provided a 63-year overlap with the tree-ring chronologies, so these three variables were chosen for the analysis. It was assumed that pressure was linked to growth by its association with general circulation of the atmosphere, frontal activity, and the movement of cyclones and anticyclones, which influenced the variables of temperature, moisture, wind, and light, variables that in turn impinge on plant processes influencing growth.

The Palmer Drought Severity Indices (Palmer 1965), which were subsequently analyzed by Stockton and Meko (1975, 1983), and the 700 millibar (mb) pressure patterns were also considered. Because both of these records began after 1930, they provided only 33 years of overlap with the tree-ring data. This overlap was considered to be too short a period for calibrating such a large spatial grid. The degrees of freedom (df) after calibration would have been insufficient for rigorous statistical tests. In addition, these data lacked independent data that could be used for verification tests.

Experience with calibrating a large spatial grid of sea-level pressure (Fritts et al. 1971; Fritts 1976) indicated that reconstruction of pressure was possible over the North Pacific Ocean and North American continent, but there was little prospect for reconstruction of this variable over the North Atlantic, at least not from the chronologies in the western United States. A new sea-level pressure grid was selected that included 96 data points between 100°E and 80°W, 20° and 70°N. It was the same grid size as before but covered more area to the west of the tree-ring grid, including parts of the East Asian mainland (see Fig. 4.4).

A. Twentieth-Century Data

1. The Pressure Data

It was noted in Chapter 2 that the tree-ring responses to climate varied from one season to the next because growth processes could be limited by a variety of environmental factors throughout the year. The sea-level pressure data were analyzed in the following manner (Blasing 1975) in the hope that the spatial variations in pressure would reflect general circulation features that are correlated with seasonal microclimatic factors limiting tree-ring growth on the sampled sites. It was assumed that whenever large-scale perturbations in climate occur and persist over one or more seasons, they will appear as large-scale anomalies in the sea-level pressure that can be calibrated with anomalies in other climatic variables (Kutzbach and Guetter 1980) as well as with tree growth.

Monthly mean sea-level pressure data from the Extended Forecast Branch, National Weather Service, National Oceanographic and Atmospheric Administration, were selected for the years 1899 to 1966 (Blasing 1975). The record was extended to June 1971 with data received from the University of East Anglia. Seventy-six grid points were selected at 10-degree intervals covering the middle latitude (20° to 50°N) regions of the North Pacific extending from 80°W to 100°E longitude. To equalize the density of the grid at 60° and 70°N and to eliminate some of the redundancy in adjacent values due to various interpretations of the observational data, 20 additional data points were selected at intervals of 20° longitude from 80°W to 100°E (Blasing 1975). This made a total of 96 points in the sea-level pressure grid. Missing grid-point values were interpolated from surrounding grid points, and the respective mean values were used for missing monthly values in December 1944 and June 1970.

At a later stage in our work a comparable data set, obtained from the National Center for Atmospheric Research, was analyzed by Trenberth and Paolino (1980). The two data sets were examined carefully over the area covered by the 96-point grid. Each 10-degree grid point was compared, and those with values differing by more than one millibar were noted (Lofgren 1978). Two months, January 1906 and November 1931, exhibited the most notable differences, with 30 and 27 grid points out of a total of 253 disagreeing by two or more millibars. Excluding these two months, there were only 185 grid points during the period from January 1899 through December 1945 which differed by as much as two or more millibars.

More discrepancies were discerned in the data beginning in January 1946 and continuing through June 1971. The pressure measurements at the 96 grid points differed by more than one millibar 10 percent of the time. The points of divergence were not equally distributed over the grid. For example, there were no differences at two grid points (70°W 20°N and 80°W 20°N, both in the Caribbean area), and differences as high as 40 percent were found for two points (100°E 20°N and 100°E 40°N, both in Southeast Asia).

Trenberth and Paolino (1980), working with the data from the National Center for Atmospheric Research, estimated that approximately 0.47 percent of the data points contained errors, with the greatest frequency of errors occurring over Asia to the west of our selected grid in the periods before 1922 and during World War II. These errors, if present in our data set, would have their greatest effect along the western border of the grid, especially the poorly covered region of Siberia. Trenberth and Paolino (1980) and Jones (1987) attributed other errors to the dubious quality of some data and to overestimates of pressure in early years at high latitudes (around 70°–80°N) along the northern boundary of our grid. An additional source of error was identified as arising from spurious trends over the Mexican highlands. If present, they could affect the pressure data at the southeastern extremities of the grid. Because the points, which appear most likely to be in error, are found along the periphery of our selected grid, it was concluded that systematic errors causing long-term variations in sea-level pressure over the Northern Hemisphere reported by Trenberth and Paolino (1980) would be minimal throughout the central portions of the 96-point grid and maximal at the borders. The Trenberth and Paolino data were obviously superior to those used in our analysis, but it was concluded that our project was too far along to begin the calibration of the sea-level pressure again.

2. Choice of the Seasons

Blasing (1975) proposed that the climate relationships might be most clearly defined if it were possible to determine the "natural" seasons. He examined the monthly sea-level pressure over the selected 96 grid points for 1899 to 1961 and agreed with Bryson and Lahey (1958) and Dzerdzeevskii (1968), who had looked at a variety of climatic data, that December through February could safely be taken as winter and that

July and August essentially comprised summer. Blasing (1975) noted significant changes in the sea-level pressure patterns between February and March, and June and July, as well as a smaller change between April and May. These changes suggested that spring could have been divided into early spring and late spring. But subsequent calibrations using a two-month early and a two-month late spring season accounted for little climatic variance. These two-month seasons were combined into a four-month spring for all subsequent work. Data from September through November were sufficiently alike to define that period as autumn.

The seasons that were selected were (1) winter: December–February, (2) spring: March–June, (3) summer: July–August, and (4) autumn: September–November. Annual values were calculated for December through November and were assigned the date of the January value. The choices of a four-month-long spring and a two-month-long summer made it difficult to relate our results to other studies using the traditional three-month seasons. Though the choice had a valid climatological justification, the calibration depended more on the tree-ring response to climatic variability than on the interrelationships between circulation patterns in the North Pacific and the North American continent. In retrospect, it probably would have been better to have chosen the seasons using the relationships between growth and climate as reflected in some kind of response-function analyses or related results (Fritts 1974).

3. Temperature and Precipitation Grids

There were 167 temperature and precipitation records from the United States and western Canada with nearly continuous data from 1901 to 1970. These data were examined and screened, and those with undesirable characteristics were deleted (DeWitt 1978). The data sets that were retained were those with (1) the longest continuous record; (2) the least number of missing data; (3) the fewest discontinuities representing changes in instrumentation, the exposure of the station, or its location; (4) a locality that was topographically most representative of the region; (5) available nineteenth-century data for use in verification; and (6) a rural rather than urban location to minimize undetected effects of urban warming.

The screening proceeded in the following manner. The monthly and annual data were plotted as a function of time. Temperature data were then separated by season and paired with nearby stations. The cumulative temperature differences within each season were calculated and plotted from one year to the next using Cumulative Temperature Analy-

Table 3.1. Stations of the 77 and 46 Temperature Grids and the 96 and 52 Precipitation Grids

Station Name	State	Temperature 77	Temperature 46	Precipitation 96	Precipitation 52	Region No.
Aberdeen	WA	67	38	83	42	1
Abilene	TX	49		73		8
Albany	NY	42		57		11
Albuquerque	NM			54	30	4
Alpena	MI	26		39		9
Amarillo	TX	50	31	74	36	6
Baker	OR	62	35	64	33	1
Bakersfield	CA			6	5	2
Banff	AL / C	72	42	93	50	5
Baton Rouge	LA	24		35		10
Birmingham	AL			1		10
Bismarck	ND	57		59		7
Blue Hill Observatory	MA			38		11
Boise	ID	17	16	26	21	3
Brownsville	TX	51		75		8
Buffalo	NY			58		9
Burlington	VT	65				11
Cairo	IL	19				9
Calgary	AL / C	73	43	94	51	5
Canon City	CO	8	12	17	16	6
Charlotte	NC	56				11
Cheyenne	WY			88	45	6
Chicago	IL			29		9
Colville	WA			84	43	1
Concord	NH	39		53		11
Dayton	OH	60		62		9
Des Moines	IA	20		31		7
Detroit	MI			40		9
Dodge City	KS	21		32		8
Durango	CO			18	17	3
Edmonton	AL / C	74	44			5
El Paso	TX	52	32	76	37	4
Elko	NV	36	23	51	28	3
Eureka	CA	3	5	7	6	2
Fargo	ND	58		60		7
Fillmore	UT	54	33			3
Flagstaff	AZ	1	1	2	1	4

Table 3.1. (continued)

Station Name	State	Station No. in Grid Temperature 77	46	Precipitation 96	52	Region No.
Fresno	CA			8	7	2
Ft. Collins	CO	9	13			6
Galveston	TX	53		77		8
Grand Junction	CO	10	14	19	18	3
Great Falls	MT	29	18			5
Hailey Ranger Station	ID	18	17			3
Havre	MT	30	19	45	24	5
Hays	KS	22		33		8
Heber	UT			78	38	3
Helena	MT	31	20	46	25	5
Huron	SD	45		68		7
Independence	CA			9	8	3
Indianapolis	IN			30		9
Jacksonville	FL	13		22		10
Kalispell	MT	32	21	47	26	1
Kamloops	BC / C	75	45	95	52	1
Knoxville	TN	47		70		10
Lakeview	OR	43	28			3
Lander	WY			89	46	5
Lewiston	ID			27	22	1
Lincoln	NE	35		49		7
Los Angeles	CA			10	9	2
Lusk	WY			90	47	5
Lynchburg	VA	66		82		11
Macon	GA	16		25		10
Madison	WI	70		87		9
Manti	UT	55	34			3
Marquette	MI	27		41		9
Memphis	TN	48		71		10
Miles City	MT			48	27	5
Minneapolis	MN			42		7
Moab	UT			79	39	3
Montrose	CO			20	19	3
Nashville	TN			72		10
Needles	CA	4	6	11	10	4
New Haven	CT	12				11
North Platte	NE			50		7
Oklahoma City	OK	61		63		8
Parkersburg	WV			86		11
Parowan	UT	63	36	80	40	3

Table 3.1. (continued)

| Station Name | State | Station No. in Grid | | | | Region No. |
| | | Temperature | | Precipitation | | |
		77	46	96	52	
Philadelphia	PA			67		11
Phoenix	AZ		2	3	2	4
Pittsburgh	PA			66		11
Poplar	MT	33	22			5
Port Angeles	WA			85	44	1
Porthill	ID			28	23	1
Rapid City	SD	46	30	69	35	5
Red Bluff	CA	5	7	12	11	2
Regina	SA / C	77		96		7
Reno	NV	37	24	52	29	3
Riverside Fire Station	CA		8	13	12	2
Roseburg	OR	44	29	65	34	1
Roswell	NM	40	26	55	31	6
Sacramento	CA	6	9	14	13	2
Salt Lake City	UT	64	37			3
San Diego	CA		10	15	14	2
Santa Fe	NM	41	27			4
Sheridan	WY			91	48	5
Shreveport	LA	25		36		10
Spokane	WA	68	39			1
Springfield	IL	28		44		8
St. Louis	MO			43		9
New Mexico State University	NM			56	32	4
Tallahassee	FL	14		23		10
Tampa	FL	15		24		10
Tooele	UT			81	41	3
Topeka	KS	23		34		8
Trinidad	CO	11	15	21	20	6
Tucson	AZ		3	4	3	4
Valentine	NE	34				7
Vancouver	BC / C	76	46			1
Visalia	CA	7	11	16	15	2
Walla Walla	WA	69	40			1
Washington	DC			37		11
Williston	ND	59		61		7
Winnemucca	NV	38	25			3
Yellowstone Park	WY	71	41	92	49	5
Yuma	AZ	2	4	5	4	4

sis. The Mann-Kendall statistic (Bradley 1976a) was also calculated and the results tested for significance.

The precipitation data were paired and submitted to a double-mass analysis described by Kohler (1949) and the Mann-Kendall statistic to identify significant breaks in the record. Changes in the slope of the plotted data were noted in both types of analysis. The history of the stations was then checked to see if stations had been relocated or if other possible causes of the discontinuity could be detected. Only stations without apparent discontinuities were selected for this analysis.

After the stations with the most reliable records were established, the spatial coverage of stations was examined, the best of these were selected, and grids of 77 temperature records and 96 precipitation records (Table 3.1) were assembled. Occasional missing data were estimated by regression on neighboring climatic records. Approximately 60 percent of the chosen stations were located in the western United States and Canada in the area of the tree-ring grid. The remaining 40 percent formed a somewhat less dense grid over the eastern United

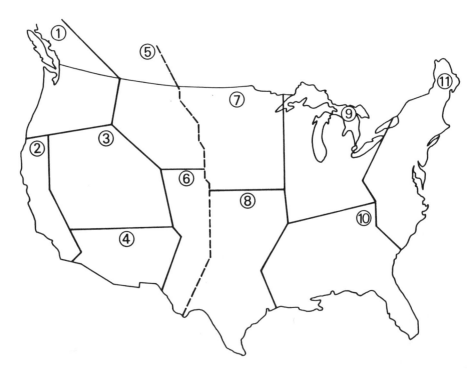

Fig. 3.1. The eleven climatic regions: (1) Columbia Basin, (2) California valleys, (3) intermontane basins, (4) Southwest deserts, (5) northern High Plains, (6) southern High Plains, (7) northern prairie, (8) southern prairie, (9) Great Lakes and Midwest, (10) Southeast, and (11) East. Regions 1–6 form the western grid.

States. These large grids provided the opportunity to examine how far climatic reconstructions could be extended beyond the area of the tree-ring grid.

Smaller subsets of 46 temperature records and 52 precipitation records (Table 3.1) were also selected from stations located within 320 km of the tree sites. These smaller sets were chosen to provide a measure of the maximum relationships that might exist. More urban stations had to be included in these smaller subsets because of gaps in the western portion of the grid. The smaller grids and the western half of the larger grids were divided into regions of relatively equal size (regions 1–6), each with a more-or-less homogeneous climate. The eastern half of the large grids was divided into regions 7–11 (Fig. 3.1, Table 3.1).

B. Nineteenth-Century Data

1. Quantity

Insufficient sea-level pressure data were available before 1899 for verification purposes, but considerable early temperature and precipita-

Fig. 3.2. The number of years with independent temperature data available before 1901. The numbers are plotted at the climatic station grid points.

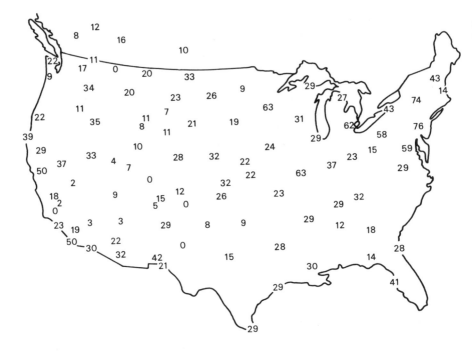

Fig. 3.3. The number of years with independent precipitation data available before 1901. The numbers are plotted at the climatic station grid points.

tion data existed. All observations before 1901 for stations included in the grids above were used for possible independent testing (Figs. 3.2 and 3.3, Table 3.1). The number of independent observations for temperature ranged from 0 to 47 years in the West and from 0 to 118 years in the East. Some stations in the southwestern deserts and Great Basin had the shortest independent records. Many of the independent data sets were discontinuous, which limited the kinds of analysis that could be used for verification. None of these early records was screened for homogeneity because too few data were available for a rigorous screening test.

2. Quality

The agency responsible for data collection in the nineteenth century changed from time to time. Each agency used the data for a different purpose and adopted different techniques for measurement and analysis (Kennedy and Gordon 1980). The National Weather Service used the average of the daily maximum and minimum temperatures to compute a daily average temperature. The Surgeon General's Office averaged four observations into a single daily value. In early records the maximum and minimum temperatures may not have been observed and recorded. Procedural changes of this kind could lead to substantial differences in the daily meteorological observations and probably contributed to errors in the monthly climatic estimates (Rumbaugh 1934).

Changes in the exposure or elevation of the instrumentation were common (Kennedy and Gordon 1980). The reliability, ease of reading and recording, optimal operating conditions, and calibration procedures varied for different instruments. Mercury and alcohol thermometers were used interchangeably (Abbe 1893), and the type of rain gauge was changed at times (U.S. Department of Commerce, Weather Bureau 1950). Thermometers were not calibrated before 1843, and calibration techniques varied after that period. These differences probably contributed to more nonclimatic variability in the nineteenth than in the twentieth century (Kennedy and Gordon 1980).

This nonclimatic variability would produce error in the verification tests, lead to statistical underestimation of the real climatic variations for this time period, and introduce a downward bias in the verification results. In spite of this bias, verification with the independent nineteenth-century record seemed appropriate since it left the 63-year-long continuous record of the twentieth century intact for calibration. Since all temperature and precipitation verification tests were based on these early data, the numerous sources of error that would affect the results should be considered in evaluating the verification tests.

C. Twentieth-Century Mean Climatic Patterns

The distributions of ocean, land, and mountains are major controls of the climatic circulation patterns over the earth's surface. Wendland and Bryson (1981) describe these patterns as particular sources and sinks that influence the flow of air (Fig. 3.4 [top], Table 3.2). According to Wendland and Bryson, the Arctic ice pack is a source of cold, dry air flowing into North America for eleven months of the year. This air often flows across the Pole into the extreme North Pacific and the North American Arctic and Subarctic, sometimes reaching far south into the United States during winter and spring. Another airstream, especially important in winter, has its source in central East Asia for a duration of eleven months. Cold continental air is transported in the westerlies from central Asia across the northern portions of the Pacific, penetrating the North American continent along the Canadian Pacific Coast. At times this airstream extends as far east and south as Washing-

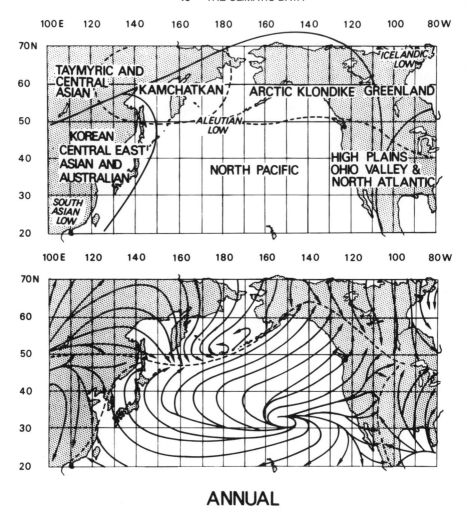

ANNUAL

Fig. 3.4. *Top*: The sources and sinks that influence the flow of air over the 96-point pressure grid used in this analysis. *Solid lines*: approximate boundaries of temperate and tropical air sources; *dashed lines*: approximate boundaries of important Arctic sources. *Bottom*: The annual surface airflow over the 96-point pressure grid. *Arrows*: approximate mean vectors of flow; *dashed lines*: average zone of confluence. Redrawn and generalized from Wendland and Bryson (1981).

ton and Oregon. However, the cold Asiatic air is modified by subsiding air and the clockwise gyre of the North Pacific anticyclone (Fig. 3.4), which is a source of milder moist air transported eastward. This air penetrates North America along a wedge pointing eastward across the north-central United States in winter and southern Canada in summer (Bryson and Hare 1974).

An airstream takes on temperature and moisture characteristics at the surface depending on the type of surface and the topography over which it passes. Thus, the air is drier and warmer in the lee of high mountains and contains more moisture near the low-level routes. For example, an offshoot of the Arctic ice-pack source, a shorter stream of subsiding air, is carried from this source eastward and then southeastward into the southwestern United States and Mexico, where uplifting by mountains removes what moisture is present and creates very arid conditions in the lee of the mountains.

A zone of confluence is found over the north-south-trending western Cordillera, which generally blocks or deflects most of the lower-level air flowing from the North Pacific as the upper portion of the airstream moves across western North America. On those occasions, when there is sufficient energy or instability, some low-level air crosses the mountain crests. According to Bryson and Hare (1974), storms and the associated airflow in the low-level westerlies can cross the Cordillera along three main routes: "(1) where the westerlies reaching the coast

Table 3.2. Northern Hemisphere Airstream Sources Ranked by Duration of Tenure

Source	Duration			Area
	Maximum (mos per year)	Months	Season of Maximum	% Northern Hemisphere Covered for 6 Months
North Pacific	12	—	summer	24
Arctic	11[a]	Jan – Nov	spring	3
Central Asian	11[a]	Jun – Apr	winter	3
Central East Asian	11[a]	May – Mar	winter	1
Greenland	9	Sep – May	spring	1
Australian	7	Mar – Sep	summer	5
Ohio Valley	6	Aug – Jan	autumn	< 1
Taymyric	4	Apr, Jul Sep, Oct	Jul	—
Korean	3	Mar – May	Mar	—
Klondike	3	Jan – Mar	Feb	—
High Plains	2	Dec – Jan	Dec	—
Kamchatkan	1	Jun	Jun	—

Source: Wendland and Bryson (1981).
[a]Probably annual.

are on the average strongest, near latitudes 45 to 50°N; (2) through the Columbia River–Snake River–Wyoming Basin gap; and (3) through the lower region along the Mexican–United States border."

A weakened flow of air enters the continent at altitudes above the level of the western plateau, then descends the eastern slopes of the mountains, becoming warmer and less humid because of adiabatic compressions (Bryson and Hare 1974). To the east of the mountains, cold, dry air can sweep southward from the frozen Arctic to as far south as the Isthmus of Tehuantepec, Mexico. Warm, moist air can sweep northward from the North Atlantic source that arises in the tropical seas (Wendland and Bryson 1981), penetrating as far northward as southern Canada in a few days' time. Tropical air rarely penetrates beyond southern Canada at the ground level. Some entrainment and mixing of the air masses can occur throughout the eastern portions of the North American continent, but because of density differences in the air masses, macroscale mixing is largely inhibited.

Synoptic-scale disturbances called *cyclones* and *anticyclones*, associated with fronts and jet streams, form moving waves in the general westerly current or in those parts of it near the east-west confluence zones (Lamb 1972; Barry and Perry 1973). According to Bryson and Hare (1974), the latitudes of greatest cyclonic disturbance over North America are 45°N in winter, 43°N in spring, 54°N in summer, and 48°N in autumn. In winter and spring the variance of the flow is high across the whole continent, whereas in summer and to some extent in autumn the interior has much less disturbed flow. Maximum disturbance activity is then over the oceans.

An anticyclonic eddy that develops over the High Plains between the westerlies to the north and the trade winds to the south is an important air source in December and January (Table 3.2). Sometimes this eddy forms from the remains of an outbreak of Arctic air; other times it is composed of milder Pacific air flowing over the Rocky Mountains. From August through January with a maximum in autumn, air with clockwise flow arises in Arizona, traverses the central Plains, and reaches the Ohio Valley (Table 3.2).

Larger, more-or-less permanent, subpolar low-pressure zones develop near the cold side of the strongest thermal gradient (Lamb 1972; Johnson 1980), where the average surface pressure is kept low by the passage of frequent cyclones. The low developing over the Aleutian Islands (Aleutian Low) is the most important of these to the area of study. A second low that is often near Iceland can cause a southward flow and instability in eastern Canada; another low is associated with the summer monsoon in Southeast Asia (Fig. 3.4, top).

A mean annual confluence zone corresponding to the average position of the Arctic Front stretches along 50°N latitude across the North

Pacific to 64°N in western Canada. This zone dips southeast to join with a confluence zone stretching from the central Plains across the Great Lakes. Southward flow of Arctic air feeds the Aleutian Low and dominates most of Canada and Alaska. Warmer Pacific air from the North Pacific High pushes into the western Cordillera as far as the eastern front of the Rocky Mountains. Easterly flow of moist air into the Gulf of Mexico is diverted northward to meet the Arctic air along the zone of confluence. The variance of this confluent zone is high, however, because the Arctic Front may extend for at least one month's duration as far south as 40°N but retreats to the forest-tundra boundary in northern Canada in the summer. In addition, the Pacific air can penetrate to the Atlantic Coast, on the average, of one month's duration and then retreat to the Cordilleran chain during the spring and summer (Wendland and Bryson 1981).

During cold months Pacific Coast precipitation is highly dependent on sea-level pressure in the eastern North Pacific (Namias 1982a). Heavy precipitation is often associated with much lower than normal pressure to the northwest of the affected area, and light or no precipitation is associated with higher than normal pressure in the eastern North Pacific. Temperatures along the coast may depend on the predominant airflow, which is linked to upwelling and sea-surface temperatures. Namias (1982a) shows that warm water off the West Coast is usually associated with cold water in the central Pacific or vice versa. During summer, upwelling usually depends on the North Pacific anticyclone, which undergoes sizable variations in position and strength from one summer to the next.

The conditions over the North Pacific may also be linked to conditions over the North American continent and the North Atlantic (Namias 1982b; Trenberth et al. 1988). Dry and warm conditions in the Great Plains are often associated with (1) strong upper-level anticyclones over the affected area, (2) well-developed anticyclones in the North Pacific and North Atlantic, and (3) a trough over the west coast of North America. These conditions sometimes can be linked to a northerly displacement of the intertropical convergence zone southeast of Hawaii (Trenberth et al. 1988). Linkages may involve the relative strengths of the southern and northern branches of the jet stream, which control the movement of storms bringing wet weather to southern California or the Columbia Basin. The northern branch can also be displaced well to the north, diverting storms northward into the Gulf of Alaska (Trenberth et al. 1988).

4

Principal Components

The information at any individual grid point is not independent of the information at other grid points, particularly those that surround it. This interdependence produces multicollinearity among the grid points of both the tree-ring chronologies and the climatic data. Multicollinearity in these data sets (Cropper 1984) can lead to (1) imprecision of regression estimates, (2) incorrect rejection of variables, and (3) indeterminance of statistical estimates (Johnston 1972). Johnston (1972, p. 207) also states: "If forecasting is a primary objective, then intercorrelation of explanatory variables may not be too serious, provided it may reasonably be expected to continue in the future."

One approach to the problem of multicollinearity extracts the eigenvectors of the spatial arrays and calculates their principal components (PCs), which are orthogonal or uncorrelated representations of the original data (Daultrey 1976). In addition, the patterns of the largest-scale and most important eigenvectors often have a physical interpretation, although the lack of an interpretation may not preclude their use in calibration if they reduce a sufficient amount of variance. Moreover, the smallest PCs, which represent noise and small-scale features in the data, can be deleted from the sample. This procedure excludes some of the small-scale variability and random variations and emphasizes the large-scale features, which are more likely to represent meaningful climatic linkages. It also reduces the number of variables and thus simplifies the structure of the model to be calibrated.

A. Principal Components of the Tree-Ring Grid

The first 10 eigenvectors of the 65-chronology grid and their corresponding PCs are shown in the order of their importance (Figs. 4.1 and 4.2). Each eigenvector portrays orthogonal patterns through space, and each principal component portrays orthogonal patterns through time. The elements (weights) of the eigenvectors (Fig. 4.1) can be contoured to reveal the spatial patterns that they represent. Each pattern should be interpreted in terms of both positive and negative representations by using the contour patterns with and without the signs reversed. Eigenvector 1 is the largest-scale pattern in tree growth, as it reduces 25 percent of the chronology variance. The most extreme (lowest) values are assigned to chronologies south of the junction of the Four Corners states. The magnitudes of the eigenvector elements decrease in all directions (become less negative) from this central point, and the elements change sign for a small number of chronologies in the northwestern United States and Canada. These are the chronologies at distances of 2,000 km or greater with negative correlation coefficients with chronologies to the southwest (Cropper and Fritts 1982). The patterns reflected in this eigenvector express a large-scale coherency between all chronologies over the southern and central portions of the map. Eigenvector 1 explains the most variance in years when the rings in the southwestern United States are either wide or narrow while the rings in the Northwest are average or opposite those to the south.

The PC for eigenvector 1 (Fig. 4.2) is low when growth is high throughout the western United States and is high when growth is low over the same area of the grid. This PC varies in a more or less random fashion during the seventeenth through nineteenth centuries, but becomes more persistent and negative after 1900, when growth increases throughout the western United States. The values of PC 1 gradually rise during the century as growth declines throughout the Southwest.

Eigenvector 2 reduces 10 percent of the chronology variance. It has large positive values for chronologies in the northwestern United States and southwestern Canada, with smaller negative values for chronologies in New Mexico and Arizona. This eigenvector portrays opposing variations between the chronologies at the northwest and southeast corners of the tree-ring chronology grid, emphasizing the seesaw pattern that produces inverse correlations between chronologies at great distances. The plot of PC 2 (Fig. 4.2) shows less variation than PC 1 and is marked by low-frequency trends in the seventeenth and twentieth centuries with highest values early in the twentieth century and lowest values beginning in the 1930s. Both PC 1 and 2 exhibit shifts to lower-frequency variations that begin at the turn of the twentieth century and are unlike the patterns that can be noted in the previous three centuries of growth.

Eigenvectors 3 through 10 show progressively more complicated spatial patterns representing smaller-scale variations in growth that may be added to eigenvectors 1 and 2 to account for the most important modes of correlation in the 65-chronology grid. The eigenvectors re-

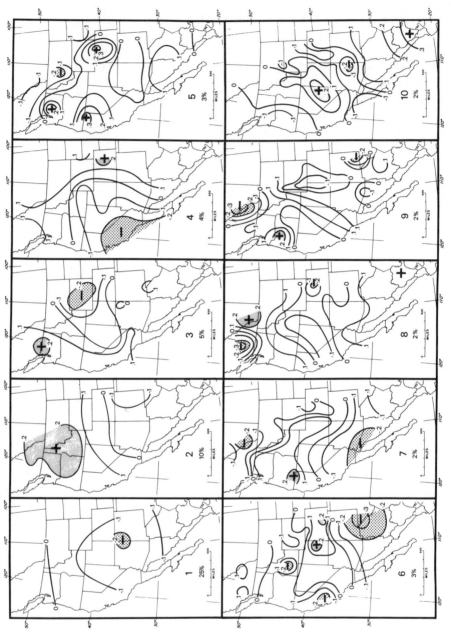

Fig. 4.1. The first 10 eigenvectors of the unmodeled tree-ring chronologies [1].

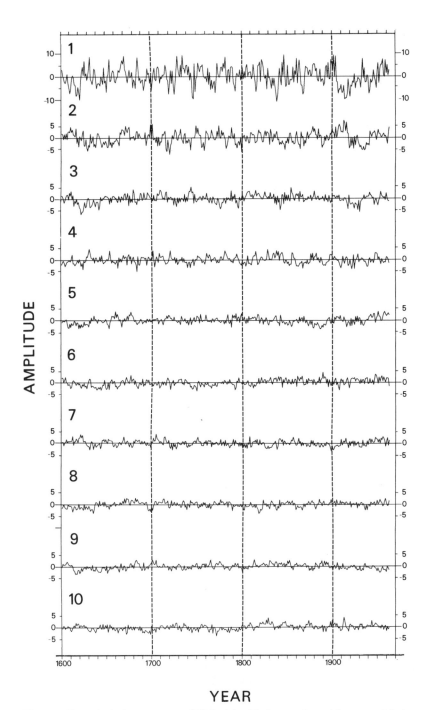

Fig. 4.2. The principal components (PCs) of the 10 eigenvectors of the unmodeled tree-ring chronologies [*I*] shown in Figure 4.1.

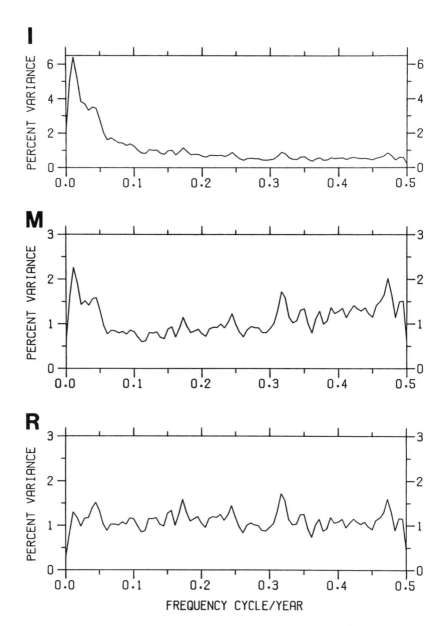

Fig. 4.3. The cumulative variance spectra of the first 15 PCs of the 65 chronologies: [*I*], before time-series modeling; [*M*], after first-order autocorrelation was removed; and [*R*], after all significant autocorrelation and moving average terms were removed (Rose 1983).

duce less and less variance, reflected in the diminishing amplitudes of the PCs. The first 10 eigenvectors account for 58 percent of the chronology variance. The first 15 eigenvectors account for 69 percent of the chronology variance.

Eigenvectors were also obtained for the 65-grid chronologies after they were corrected for the first-order autocorrelation (Eq. 2.3). The eigenvectors of this set resemble those of the uncorrected chronology set, and the first 10 eigenvectors of both sets are well correlated, although the signs of the coefficients are sometimes reversed. The first 10 eigenvectors for the corrected set account for 59 percent of the chronology variance. The first 15 eigenvectors account for 67 percent of the chronology variance. The uncorrected and corrected growth data sets appear to have very similar patterns.

Later in the study, power spectra (Rose 1983) were calculated for the PCs of growth, and the spectral estimates were summed over all 15 PCs to evaluate differences in the frequency responses of the two data sets (Fig. 4.3, I and M). The untransformed data (I) had the characteristic red-noise spectrum. The transformations using only the first order autocorrelation (M) had shifted the variances too much to the right so that there was less intermediate-frequency variance than one would expect from a white-noise spectrum. This difficulty was overcome by applying higher-order terms with the methods of Box and Jenkins (1976) (Fig. 4.3, R), which reduced the data to a truly "white noise" data set. The newer data did affect some details of the reconstructed maps and plots, but the only major difference was visible in the spectra of the different data sets. It was not possible to repeat all of the work, however, so the results of the I and M data sets (see Chap. 5, Sec. B) are used throughout the book.

The means and variances for the PCs of growth were calculated over the independent and dependent time periods and compared in Table 4.1 to examine the following question: Are there any large growth differences apparent between these two time periods which could indicate differences in climate? The total fractional variances shown at the bottom of Table 4.1 were 1.11 and 1.10 for the I and M data sets, respectively; indicating that the twentieth-century chronology variance was 110 to 111 percent of the variance in the seventeenth through nineteenth centuries. Data for the individual PCs reveal that the twentieth-century variance increase, noted in Figure 4.2, is largely attributable to changes in the first two PCs of the chronology variance, with variance changes of 129 and 167 percent. This is a major increase in growth variance, one that is likely to be present in estimates of climate. However, it is possible that the reconstructed variance in the twentieth century could be less than before if the transfer-function

Table 4.1. PC Variances, Means, and Fractional Variance for the *I* and *M* Data Sets from the 65-Grid

Component Number	With Autocorrelation (*I*)				Autocorrelation Removed (*M*)			
	1901–1962		1602–1900	Fractional Variance	1901–1962		1602–1900	Fractional Variance
	Variance	Mean	Variance + Square of Means[a]	1901–1962 / 1601–1900	Variance	Mean	Variance + Square of Means[a]	1901–1962 / 1601–1900
1	21.79	−1.31	16.90	1.29	19.02	−0.87	15.95	1.19
2	10.02	0.31	6.01	1.67	7.63	0.20	6.16	1.23
3	3.32	−0.49	3.67	0.90	3.28	0.31	3.79	0.87
4	2.52	−0.33	2.83	0.89	2.80	−0.19	2.69	1.04
5	2.68	0.61	2.36	1.14	3.07	0.04	2.00	1.54
6	1.75	0.78	2.51	0.70	1.77	0.41	2.04	0.87
7	1.97	0.18	1.61	1.22	2.29	0.02	1.72	1.33
8	0.98	0.32	1.74	0.56	1.12	0.03	1.55	0.72
9	0.84	−0.69	1.91	0.44	1.20	0.32	1.46	0.82
10	1.23	0.22	1.44	0.85	1.26	0.24	1.40	0.90
11	0.77	0.51	1.63	0.47	0.94	−0.49	1.48	0.64
12	1.06	−0.26	1.30	0.82	1.36	0.17	1.14	1.19
13	1.18	0.32	1.16	1.02	1.01	−0.14	1.12	0.90
14	1.13	−0.34	1.15	0.98	1.07	−0.05	0.98	1.09
15	1.23	0.31	1.06	1.16	0.93	−0.28	1.02	0.91
Total	52.46	0.14	47.28	1.11	48.75	−0.28	44.50	1.10

Note: The variance for 1901–1962 is divided by the variance for 1601–1900 to show the fractional variance change between the independent and dependent data sets.

[a]The squares of both the dependent and independent means of the departures from the 1900–1970 means were combined with the variance of the independent data so that they correspond to the variance changes that would be estimated from each PC.

coefficients give enough emphasis to the relationships for PC 3, 4, 6, 8, 9, 10, 11, 12, and 14, which had lower twentieth-century variance.

B. Principal Components of Climate

The eigenvectors and their PCs for the climatic data were calculated according to Equations A.1 through A.4, except that the covariance

Table 4.2. Decade Averages of PCs 1–3 for Annual Sea-level Pressure, Temperature, and Precipitation Instrumental Records in the Twentieth Century

Decade	PC 1	PC 2	PC 3
Pressure			
1901–1910	−9.39[a]	−2.69	−5.08[a]
1911–1920	−6.69[a]	−1.00	3.77[a]
1921–1930	0.82	−1.91	3.53[a]
1931–1940	4.99[a]	−0.40	1.01
1941–1950	4.29[a]	2.77	−0.56
1951–1960	0.57	1.79	−0.58
1961–1970	4.32[a]	1.46	−2.28[a]
Temperature			
1901–1910	2.41[a]	1.83	0.18
1911–1920	4.59[a]	−0.12	1.49
1921–1930	−0.62	−0.90	0.11
1931–1940	−5.87[a]	−1.05	−0.47
1941–1950	−0.71	−1.88[a]	−0.28
1951–1960	−1.03	−0.99	−1.07
1961–1970	1.23	3.10[a]	0.02
Precipitation			
1901–1910	−1.02	−0.33	1.25
1911–1920	−1.08	−0.64	−0.29
1921–1930	−0.84	0.69	−1.75
1931–1940	2.33[a]	−1.30	−1.21
1941–1950	−1.77	1.60	0.21
1951–1960	2.25	0.35	1.66
1961–1970	0.13	−0.37	0.12

[a]Significant at $p \geq 0.95$.

Table 4.3. The Percentage of Variance Reduced by the Most
Important Eigenvectors of the Annual Instrumental Data

Variable	\multicolumn{5}{c}{Eigenvector Number}					\multicolumn{3}{c}{Cumulative Values}		
	1	2	3	4	5	1–5	1–10	1–15
Pressure	31	14	11	8	6	70	86	92
Temperature	33	20	13	6	5	77	87	92
Precipitation	14	9	8	6	5	42	58	69

Note: Grids were 96 points for pressure, 77 stations for temperature, and 96 stations for precipitation.

rather than correlation matrix, $_mC_m$, was used for the sea-level pressure grid. The calculations were based on data from 1899 to 1970 for pressure and from 1901 to 1970 for temperature and precipitation. The eigenvectors were then converted to PCs, primarily for use in calibration and spectral analysis (Tables 4.2 and 4.3).

1. Annual Pressure

The patterns in the first 3 eigenvectors of annual average sea-level pressure and the associated temperature and precipitation patterns are shown in Figure 4.4. Eigenvector 1 reduces 31 percent of the annual sea-level pressure variance. The eigenvector is generally negative throughout the area of the map and portrays years of consistently low or high pressure over most of the grid. The positive representation of the pattern is significant and dominant for the decades of 1931–1940, 1941–1950, and 1961–1970, which were especially warm throughout the West (Table 4.2; see also Figs. 7.1 and 7.15). This eigenvector pattern and its PC are associated with declining sea-level pressure (rising PC values) noted in the twentieth century, primarily over the North Pacific and North American Arctic (see Fig. 7.8).

The negative representation of eigenvector 1 dominated sea-level pressure patterns in 1901–1920 (Table 4.2), a cool period in the North American West (see Figs. 7.1 and 7.15 and Boden et al. 1990). This anomaly appears to be associated with a strengthened North Pacific and Klondike High and a weakened Aleutian Low. This condition created an anomalous flow in the eastern North Pacific with more advection of cool air from the Arctic. Fewer storms were diverted to the extreme Southwest, perhaps with more storm activity in the Pacific Northwest, and more precipitation along the United States and Canadian border (see Fig. 7.4).

As the values of PC 1 increased from the 1920s to 1930s, the Aleutian Low strengthened and became southeasterly displaced, and the Klondike High weakened. More storms may have moved through the Canadian Arctic. An enhanced southward airflow in the central North Pacific and a northward airflow along the California coast may have caused some Pacific storms to travel farther south than normal, bringing enhanced moisture and warmth to the Southwest and somewhat drier conditions to the Northwest (see Fig. 7.8).

Eigenvector 2 reduces 14 percent of the sea-level pressure variance, but none of the PC averages in the twentieth century is significant. The positive representation of this eigenvector is associated with higher than average pressure over the eastern and central North Pacific and lower than average pressure over the North American and Siberian Arctic coasts. This representation appears to be a blocking pattern with reduced storm activity in the Southwest and enhanced storm activity in the Arctic. The clockwise flow around the anomalous high brings cool, dry air to the Great Basin and Plains states. Areas of above-average precipitation with inferred enhanced storm activity can be noted along the northwestern United States and Canadian border and from the central Plains to the Atlantic Coast.

In its negative representation eigenvector 2 exhibits a low-pressure anomaly over the North Pacific with high-pressure anomalies over the North American and Siberian Arctic. Also, it appears to represent a strengthened and displaced Aleutian Low, a weakened North Pacific High, or both. The southward flow of cold air from Alaska and western Canada would be reduced, a development associated with warmer temperatures and perhaps higher precipitation in the West.

Eigenvector 3 reduces 11 percent of the sea-level pressure variance. Its positive representation amounts to a negative pressure anomaly over Alaska and the western Canadian Arctic. A corresponding weak anomaly of similar sign occurs in the southwestern quadrant of the map, and anomalies of the opposite sign can be noted in the quadrants to the west and south. The decade averages of the PCs (Table 4.2) are significant and positive in 1911–1920 and 1921–1930. This pattern may be associated with a blocking high-pressure ridge near the United States coast. Cool and dry conditions in the West and warmth in the East may be a result of southward airflow in the West, reduced storm activity from the block, and an increase of storms moving from the Canadian West to the United States Atlantic Coast.

ANNUAL PRESSURE

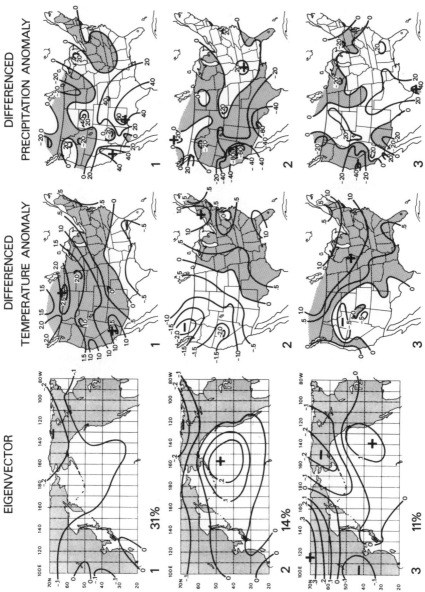

Fig. 4.4. The first 3 eigenvectors of annual pressure with the variance reduced (*left*) and the associated anomalies in temperature and precipitation (*center and right*). The last two are obtained by calculating the differences between the average of the five anomaly patterns for the 5 years of lowest-pressure PCs and the average of the five anomaly patterns for 5 years of the highest-pressure PCs. *Shading*: high anomalies of temperature and low anomalies of precipitation.

ANNUAL

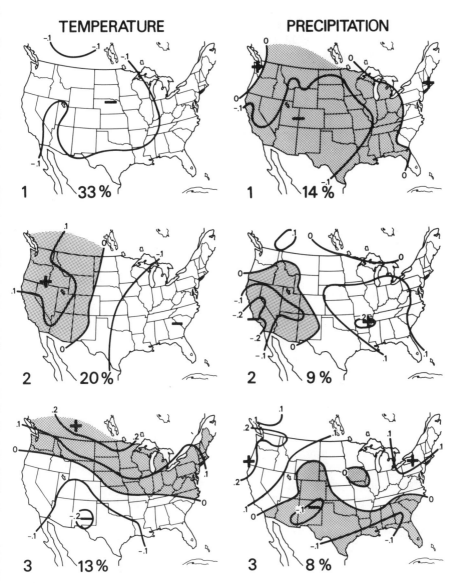

Fig. 4.5. The first 3 eigenvectors of annual temperature and precipitation with the percentage of variance that they reduce. *Shading*: high anomalies of temperature and low anomalies of precipitation.

The negative representation of eigenvector 3 may indicate a high-pressure anomaly over Alaska and an anomalous low off the United States Pacific Coast. The PCs have a significant and negative average in 1901–1910 and 1961–1970. The frequency or magnitude of storm activity appears to increase in the West with precipitation much above average for the far western United States and temperatures cooler from Washington to the East Coast.

2. Annual Temperature and Precipitation

The percentages of variance reduced by the first few eigenvectors of annual temperature (Table 4.3) are generally larger than those for pressure, and those for precipitation are smaller. The first 10 annual precipitation eigenvectors reduce only 58 percent of the variance, a value comparable to the 59 percent of the variance reduced by the first 10 eigenvectors of tree growth. The higher percentages for temperature and pressure suggest that the patterns of variation are larger in scale and can be represented by a smaller number of eigenvectors than those for precipitation and tree growth.

One possible explanation for the small-scale variations in tree growth is that the major limiting factor to growth on these sites is precipitation, and it exhibits small-scale variations over space. Also, it is possible that differences between species and sites in the response to seasonal variations in climate contribute to small-scale differences within the tree-ring grid. Our calibrations and verifications show that the large-scale features of temperature were better reconstructed from this tree-ring grid than large-scale features of precipitation. If temperature is as important to growth as precipitation on these sites and if temperature exhibits larger-scale variations than precipitation, then one would expect growth to vary in spatial scales intermediate to those of precipitation and temperature. Since growth variations are as small as or smaller than those for precipitation, the varying responses between different chronologies appear to be the better explanation than the high limitation of precipitation, although both probably contribute to small-scale variations in growth.

The first 3 eigenvectors of annual temperature shown in Figure 4.5 reduce 66 of the variance, but the first 3 eigenvectors of precipitation reduce only 31 percent of the variance (Table 4.3). The decade averages for the temperature PCs during the twentieth century are more often significant than those for precipitation (Table 4.2), perhaps because large-scale temperature trends are more apparent. The elements of the first eigenvector (Fig. 4.5) for temperature are negative throughout the map. All but those along the East and Northwest coasts are negative for precipitation. The first few eigenvectors are associated with a predominance of either high or low anomalies for large areas of the grid. The decade averages for PC 1 of annual temperature are positive and significant in 1901–1920 and positive but insignificant in 1961–1970, representing relatively cool decades throughout the grid (Table 4.2, also see Fig. 7.1). They are negative and significant for 1931–1940 and insignificant for 1951–1960, indicating overall warmth. The decade averages for PC 1 of annual precipitation are positive and significant in 1931–1940 and positive and insignificant in 1951–1960, when precipitation was low throughout the grid. The averages of the PCs are low in 1901–1920 and 1941–1950, when wetter conditions existed.

The second eigenvectors for both variables portray more east-west variations, with nodes running through the High Plains or along the Rocky Mountain crest. The decade averages of PC 2 for temperature are positive in 1901–1910 and 1961–1970, the latter period being significant when the West was warmer than the East (see Fig. 7.1). The averages are negative for other decades, indicating cooler conditions in the West, but only in 1941–1950 is the average significant (see Figs. 7.1 and 7.15). Three decade averages of PC 2 for precipitation are negative and four are positive, but none are significant. East-west variations in precipitation occurred but are not as prominent.

Eigenvector 3 portrays north-south variations in both temperature and precipitation, but no decade averages of the PCs are significant. These eigenvector patterns are similar to those of Walsh and Mostek (1980) in that the first eigenvector of each variable contains anomalies of the same sign over most of the United States; the second and third modes describe gradients in approximately perpendicular directions.

No more than the first 15 PCs of sea-level pressure, temperature, and tree growth, and no more than the first 20 PCs of precipitation were used in the following analysis. All but the number of PCs of tree growth were varied to obtain the optimum calibration models for reconstructing climate (see App. 1, Sec. G, for more discussion on the number of PCs used in the modeling work).

5

Calibration, Verification, and Merging

A. Calibration Models

Biological evidence suggests that different climatic factors can become limiting to growth in different seasons and that the effects of these factors on ring width can lag one or more years behind the original limiting conditions of climate (Fritts 1976). Growth may be directly or inversely related to these limiting factors, and autocorrelation is important since the growth in any one year may be dependent on the growth in prior years as well as on the prior year's and current year's variations in climate (Fig. 5.1).

Many nonlinear relationships have been noted between hourly changes in environmental conditions and processes limiting tree growth (Kramer and Kozlowski 1979). Nevertheless, the ring width is the integral of these processes averaged over many hours, days, and weeks throughout the living tissues of the tree. Dendrochronological procedures integrate these values further by averaging the variations over two radii from many trees growing on the selected site. Calibration procedures also average the effects over 60 years, over the 65 tree-ring chronologies, and up to 96 climatic data points used in this work. With this much averaging, many of the complex nonlinear relationships were reduced to a linear form in the averaged result.

It could be argued that one should use nonlinear models in this type of analysis. Nonlinear models, however, often consume more degrees of freedom (df). Since only 63 years were available for calibration, and spatial variations with lags and leads had to be modeled, it was impractical to fit nonlinear models at this stage in the analysis. Still, nonlinear models could reveal interesting interactions and should be used to analyze the climatic relationships for individual chronologies in

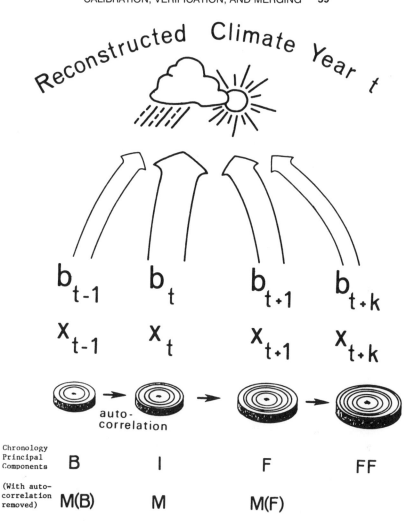

Fig. 5.1. Linear coefficients of a transfer function from b_{t-1} to b_{t+k} are multiplied by ring width or other types of ring data x_{t-1} to x_{t+k} and the products summed to obtain an estimate of year t climate. In this particular study, matrices of the coefficients are multiplied by corresponding matrices of PCs of ring width B, I, F, and FF or $M(B)$, M, and $M(F)$ depending upon whether first-order autocorrelation was removed.

a particular region where there is less averaging and an adequate chronology signal of climate is present (Graumlich and Brubaker 1987; Graumlich 1991; Fritts et al. 1991).

The term *model* refers to one particular transfer-function equation that can be applied to tree-ring data to make a climatic reconstruction

(Fritts et al. 1971; Webb and Clarke 1977). The models were simplified somewhat by calibrating only one climatic variable for one season at a time. This simplification seemed appropriate because of the marked differences that were noted in response functions from one season to the next. Models of different structure were used to assess differences in response among the 65 chronologies, spatial variations, lags in the relationship, and autocorrelation in the tree-ring data set.

The objective, at this point in the investigation, was to examine a variety of seasonal models with different structures and to use the calibration and verification statistics to decide which models provided the "best" estimates of climate. The results of the best two or three models for each variable and season were combined and statistics of the combinations used to determine whether there was improvement. Finally, the combinations with the best statistics for each season were combined to obtain annual climatic estimates.

A linear equation was obtained that relates the variations in the predictand variables (the climatic variables to be reconstructed) to the variations in one or more predictor variables (the standardized tree-ring chronologies). The coefficients of the transfer function were uniquely determined by and were dependent on the information used for calibration. Canonical regression (Glahn 1968; Blasing 1975, 1978) was used to relate statistically a number of PCs of the 65 tree-growth chronologies to a number of PCs of climate (Fritts 1976). Different lags and leads were used to represent lags and leads in the important biological and autoregressive relationships that influence growth. The PCs were used to simplify the problem, to help preserve available df, to reduce the colinearity in the system, and to focus on the largest-scale variations. There were simply too many variables (grid points) to solve the problem directly.

The coefficients of the canonical regression were used to transfer spatial variations in tree growth to spatial variations in climate. The canonical variates are orthogonal (uncorrelated) so that the effects of each canonical variate in regression can be tested, and those that did not contribute significant information could be eliminated without affecting the other variables in the analysis. The coefficients were estimated by least-squares techniques using data from the calibration period. The chronology indices (or PCs) for each year were multiplied by the associated coefficient and the products summed to obtain statistical estimates or reconstructions of year t climate. Each coefficient b of a transfer function (Fig. 5.1) was empirically derived, so it could vary in value and sign for each lag, each variable in regression, each calibration, each chronology set, and each climatic station in the grid.

B. Model Terminology

At the very beginning of the analysis different numbers of the largest chronology PCs were calibrated. The calibrated variance and other statistics consistently improved as PC numbers increased from 2 to 15. Thus, all 15 of the largest chronology PCs were used as the statistical predictors of climate. The 15 PCs were extracted from the chronology data using the correlation matrix and were then arranged to represent different lags and leads in the tree-ring chronology relationships (Fig. 5.1). For example, the chronology PCs for year t, the year immediately preceding and including growth, were designated as matrix I (immediate). The PCs for $(t - 1)$, the year prior to growth, were designated as matrix B (before). Matrices F and FF (following) included the PCs for the years $t + 1$ and $t + 2$, respectively, following the year of growth. The PCs for chronologies with autocorrelation removed for year t were designated as M (modeled). Those with autocorrelation removed for year $t - 1$ and $t + 1$ were $M(B)$ and $M(F)$, respectively. PCs of chronologies with autocorrelation removed for year $t + 2$ were not used. Lags greater than 2 or less than -1 were not considered.

The aforementioned notations designate the predictor structure and data type for the particular model. Models with B, I, F, and FF correspond to successive lags in the predictor PCs derived from chronologies before removing autocorrelation. Models with BM, M, MF, and MFF are similar, except that the PCs were extracted after removing the autocorrelation. The length of the calibration period was 61 to 63 years. To preserve at least 30 df for the regression analysis, no more than two sets of the chronology PCs (15 PCs per set) were used for any one calibration. Models with 15 PCs, B, I, F, FF, M, were called *singlets*, and models with 15 PCs from two different sets, such as BI, IF, IM, and FFF, were called *couplets*.

The first matrix in a model name designates the seasonal climatic variable that was reconstructed, with T representing temperature; R, precipitation (rainfall); and P, sea-level pressure. Each name begins with the number of q predictand PCs that were used as candidate predictands in the calibration. The actual number of canonical variates that were selected was always a smaller number. That helped to preserve 40 to 50 or more df in the residuals after the effects of autocorrelation of the residuals were taken into account. Although this important number was not used in the model name, it is listed in the summary tables for the chosen models included in Chapter 6.

Early experimentation included different numbers of climate PCs and different kinds of predictor singlets and couplets. Widely different model structures sometimes yielded high-quality reconstructions with distinctly different statistics, suggesting that different tree-ring characteristics may have been calibrated by the models (see App. 1, Sec. F.2). In some cases the reconstructions from models of different structures were quite different. This result was thought to reflect competing tree-growth responses between different species and contrasting sites that gave rise to the spatial variability noted earlier in the tree-ring data set. The reconstructions of two or three models showing different responses, but with high-ranking statistics, were averaged in an attempt to make use of some of the diversity in response.

Canonical regression finds the best-fitting line through the cluster of points representing the predictand data. The success of the regression is determined by calculating the percentage of the actual data variance that is mimicked by the estimate. These results are often expressed as a fraction of the total variance reduced by the reconstruction (see App. 1, Secs. B and C, for examples and more information on calibration procedures).

C. Verification

The results of transfer-function relationships modeled in the dependent period are assumed to be the same as those modeled during the independent period (Webb and Clarke 1977; Bryson 1985). The coefficients of the transfer function should be stable; that is, they should be the same no matter what data are used as the dependent data. A successful model does not simply mimic the dependent data, but also expresses a universal property regarding the tree-growth response to variations in both present and past climate. Nevertheless, the process used to optimize the coefficients of the transfer function virtually ensures that the model will be more accurate for the dependent data than for any other body of data to which it may be applied (Larson 1931; Wherry 1931; Anderson et al. 1972; Stone 1974). Thus, the predictive power of a regression model must decrease when the model is applied to independent data. This deterioration in accuracy should be measured whenever possible and the measurements used either to evaluate the performance of the model or to provide the proper perspective with which to view the climate reconstructions.

Any procedure that is used to assess the reliability of a reconstruction on independent data is a *verification procedure*. Reliability can be measured by *verification statistics* that assess the degree of association between the estimates of climate independent of the calibration data and the corresponding instrumental data they are supposed to mimic.

Not only must a successful reconstruction have significant calibration statistics, but its verification statistics must demonstrate that the independent estimates continue to be accurate and that the accuracy measurement is not likely to be the result of chance. The following discussion is an abbreviated version of an unpublished report on the topic by Gordon (1980) (more details are given in App. 1, Sec. D).

D. Verification Statistics

Statistical verification involves the comparison of independent predictand estimates with corresponding instrumental data and the calculation of a score or statistic that measures their similarity. Various statistics can be used. Some, called *parametric statistics*, involve assumptions about the underlying distributions of the data and may be sensitive to violations of those assumptions (Graumlich 1985). Others, called *nonparametric statistics*, do not involve such assumptions but are generally less sensitive measures of agreement between the predictand estimates and instrumental data. Both kinds of statistics are used in this study, and a variety of statistical tests are applied in different ways to assess different attributes of similarity.

The well-known *product moment correlation* coefficient, r (Clark 1975), measures the relative variation (covariance) that is common to two data sets and reflects the entire spectrum of variation, including both high and low frequencies. The correlation coefficient can be affected markedly by any trends in the two time series that are being compared. It is totally insensitive to differences in the scale and to differences in the mean between the two data sets. The effect of trends can be eliminated by calculating a new correlation coefficient from the first differences of the two data sets. Correlations calculated from the first differences, r_d, measure only the high-frequency variation in common expressed by the year-to-year differences.

The *sign test* is a nonparametric statistic involving a count of the number of times that departures from the sample means agree or disagree. The number of signs is significant whenever it exceeds the number expected from random numbers. The test measures the associations between two series at all frequencies but does not assume that the data are normally distributed. A similar test is made for the first differences, and the test of significance is the same as the test of departures. The sign of the first difference measures the associations at high frequencies.

The *product means* (*PM*) test (Fritts 1976) accounts for both the signs and the magnitudes of the similarities in two data sets. It emphasizes the larger deviations from the mean over the smaller deviations by collecting the products of the deviations in two separate groups based on their signs. The means of these two groups are calculated, and the difference between the absolute values of the two means is tested for significance.

The *reduction-of-error* (*RE*) statistic provides a sensitive measure of reliability and has useful diagnostic capabilities (Gordon 1980). It is similar in some respects to the explained variance statistic obtained with the calibration of the dependent data (Lorenz 1956, 1977). The value of *RE* can range from negative infinity to a maximum value of 1.0, which indicates perfect estimation. The theoretical distribution of the *RE* statistic has not been determined adequately, so its significance cannot be tested. Any positive value of *RE* indicates that the regression model, on the average, has some skill and that the reconstruction made with the particular model is of some value. The errors are unbounded, however, so that one extreme error value in what was otherwise a nearly correct set of estimates could cause the *RE* statistic to be negative.

The *RE* can be partitioned into three component parts—the *RISK*, *BIAS*, and *COVAR* terms—which express various attributes of the relationship (Gordon and LeDuc 1981; Gordon 1980). These components can be extremely useful as diagnostic tools for analyzing sources of error affecting a particular climatic reconstruction.

The *RISK* term is always negative; its absolute magnitude is a comparative measure of the variability of both the estimates and the actual observations used in testing. This term represents the risk that the model takes in making the independent estimates. Ideally, the *RISK* term should be -1.0 (see App. 1, Sec. D.4). Estimates with a small explained variance usually have *RISK* values between -0.5 and 0.0, and reconstructions that have a larger variance than the actual data will have values that are less than -1.0. This overspecification of the variance can occur when an excessive number of predictors is included in the transfer function.

The *BIAS* term is positive when the mean of the estimates is on the same side of the calibration mean as the actual independent climatic data used for the verification testing. Usually, shifts in the mean are insignificant, but for a small sample the *BIAS* can be an important *RE* component. The covariation term, *COVAR*, reflects the strength of the correlation between the reconstruction and the instrumental data and measures the similarity of the temporal patterns in these two data sets. To obtain a positive *RE*, the *RISK* term must be offset by the accuracy of the estimates as indicated by the *BIAS* and *COVAR* terms.

The partitioned *RE* components can be used in the following ways to diagnose reconstruction characteristics. Some reconstructions can successfully duplicate the temporal patterns of variation in the actual observations but contain no appreciable amount of variability. The correlation coefficient would not differentiate such a reconstruction from one with more variability, but the *RISK* term would clearly reveal this difference. Cases frequently arise in which regression estimates have a negative *RE* statistic and yet still pass a majority of other verification tests, especially the correlation statistics. In this situation, a *RISK* term that is less than -0.1 may reveal that the model has overestimated the instrumental data variance. A negative *BIAS* term may indicate differences in the reconstructed and instrumental mean values. Of course, a small *COVAR* term would occur only if there is little correlation.

It is possible that nonlinear relationships could exist between the two samples which cannot be properly evaluated by using the aforementioned verification statistics. Therefore, a *contingency analysis* (Beyer 1968) was used to test for a relationship without making any assumption about linearity. A chi-square statistic is then used to test whether the relationship is sufficiently strong to be significant.

E. Strategy Imposed by Data Availability

All available temperature and precipitation data from the same stations used for calibration but reported for years before 1901 were used for the verification of temperature and precipitation. Each statistic was calculated only for stations with at least seven independent observations of climate. A different calibration and verification strategy, called *subsample replication* (Mosteller and Tukey 1968, 1977; Stone 1974; McCarthy 1976; Gordon 1980), was used to obtain the independent data for verification of sea-level pressure (see App. 1, Sec. G.4, for more information).

All of the verification tests applied to temperature and precipitation estimates at individual stations could be applied to the gridded sea-level pressure estimates. It was more efficient, however, to apply them to the PCs of sea-level pressure, and this method avoided problems with spatial correlation. These series diminished in variance as the order of the PC increased, though, and it was not appropriate to normalize the PC values before calculating the statistics. As a result, the statistics requiring normalized estimates such as the contingency analysis could not be applied to the PCs of pressure. Differences in the structure of sea-level pressure models were evaluated by comparing the calibration and verification statistics for the estimated PCs.

F. Verification of Temperature and Precipitation

All available temperature and precipitation data (including discontinuous records) before 1901 (Figs. 3.2 and 3.3) were used for verification testing. All data were normalized using the mean and standard deviations of the calibration period. Then the six verification statistics, excluding the contingency analysis, were calculated for all grid points having at least seven independent climatic observations. The significance of each statistic was tested at the 0.95 probability level, and the percentage of tests that were significant for each of the statistics was tabulated. For the *RE* statistic, the percentage of stations with a positive value was recorded. In addition, the percentage of all tests that were significant (excluding *RE*) was tabulated.

The contingency analysis was applied to the data pooled over all stations. Other statistics that were calculated are more fully described in Appendix 1, Section D. The threshold percentages (Livezey and Chen 1983) of grid points needed for significance of the collective set of statistics were (1) 17.5 and 1.5 for the sea-level pressure data and their reconstructions, (2) 15.5 and 0.0 for the temperature data and their reconstructions, and (3) 27.5 and 0.0 for the precipitation data and their reconstructions. The low critical threshold values for the reconstructions suggested that spatial correlation in the reconstructions was not a major problem. The statistics for the final chosen models were well above the expected levels calculated from the instrumental data.

Whenever the percentage of verification tests passing was equal to or less than 5 percent or was less than the threshold values listed above, the model was considered unverified and was dropped from the analysis. The same verification statistics, as well as some calibration statistics, were then used to decide which of the verified models provided the most reliable reconstructions.

G. Verification of Sea-Level Pressure

The subsample replication of sea-level pressure (Mosteller and Tukey 1968, 1977; Stone 1974; McCarthy 1976; Gordon 1980) proceeded as

Table 5.1. Subsamples Used for Sea-level Pressure Calibration
and Verification

			Number of Years	
Partition	Source of Calibration Data[a]	Independent Estimates	Calibration	Independent
1	1899–1900 and 1916–1963	1901–1915	50	15
2	1899–1910 and 1926–1963	1911–1925	50	15
3	1899–1920 and 1936–1963	1921–1935	50	15
4	1899–1930 and 1946–1963	1931–1945	50	15
5	1899–1940 and 1956–1963	1941–1955	50	15
6	1899–1950	1951–1963[a]	52	13

[a]The 5-year overlap is used to adjust for differences in the mean, and the remaining 10 years in the subperiods are joined to obtain a continuous set of independent data for 1901 – 1963. The last year in subperiod 6 is a function of the model structure and the season of the reconstruction. Other possible last years are 1960, 1961, or 1962.

follows: the data from 1899–1963 were divided into the six subsamples (Table 5.1). In the case of the first subsample, the years 1899–1900 and 1916–1963 were used for calibration, which left years 1901–1915 for the independent verification testing. In the case of the second subsample, 1899–1910 and 1926–1963 were used for calibration, and the years 1911–1925 were left for verification. The last 5 years of the independent data in each subsample, 1–5, overlap the first 5 years in the independent subsample that follows them. Starting with subsample 2, the difference in the mean of the last 5 years of the preceding subsample and the first 5 years of subsample 2 was added to all fifteen estimates of the subsample to correct for any mean differences introduced by the subsample calibration. These corrections were made and accumulated over all six subsamples; the last 5 years of independent data for each subsample 1–5 were discarded; the remaining data were combined to form a continuous set of independent data spanning the period 1901–1963. These data were corrected for the arbitrary use of subsample 1 as a reference point by adding the difference between the mean of the instrumental data and the mean of the estimated sea-level pressure estimates. This procedure removed any differences in the means for the dependent and independent data, which would cause the *BIAS* term of the *RE* statistic always to have low values. Thus, no estimate of the accuracy in the long-term mean could be obtained because of this particular strategy.

H. Further Model Development

A large number of singlet and couplet models was examined, and those with nonsignificant statistics were eliminated from further consideration. Often significant models of widely different structure had similar and relatively high calibration and verification statistics. Bates and Granger (1969) show that the reliability of a statistical forecast may be improved by averaging statistical estimates from two or more calibrations using different combinations of predictor variables. One to three reconstructions from the highest ranking models were averaged, and new statistics for the averaged estimates were calculated by comparing the averages to the appropriate climatic data used for calibration and verification.

1. Varying the Number of Principal Components

Preliminary analysis using different numbers of both predictor and predictand PCs indicated that the variance calibrated by a model of a given structure generally increased as the number of predictand PCs used in the regression increased. Unexplained departures from this pattern were common, however, in which there were marked increases or decreases in variance calibrated associated with a change in the number of predictand PCs used in the analysis. These changes appeared to be related to differences in the number and size of the canonical variates that were retained in the analyses and depended on the particular kind of climatic relationship that was calibrated. These changes were assumed to reflect real differences in the correlations between particular spatial variations in growth and climate, so the following strategy was adopted. A total of 15 and 30 predictor PCs of the tree-ring chronologies were used in all singlet and couplet models, respectively, for temperature and precipitation, but the number of predictand PCs was varied from 2 to some maximum value.

A variety of methods was used to identify the noise levels of the eigenvectors and the PCs (see App. 1, Sec. G.1). Some methods were more stringent than others, but there was no general agreement among the methods or among the results as to the optimum number to use in these calibrations. Rather than singling out one method, it seemed best to start with a fairly high number of eigenvectors based on the least stringent criteria and then apply more stringent criteria as the models became more highly developed. The F statistic eliminated all canonical variates that did not contribute significantly to the regression, so elimination of PCs before analysis was not particularly critical. A brief description is presented in the following section, and more details are included in Appendix 1, Section G.

The first models used the largest 15 PCs of the 65 chronologies to reconstruct 15 PCs of temperature or 20 PCs of precipitation. The canonical regression selection procedure eliminated the nonsignificant canonical variates, which helped to preserve adequate df in the model equation. Eleven combinations of couplet models and the I and F singlets (see Table A.6) were calibrated with each variable, season, and grid. Seven calibration statistics and seventeen verification statistics were computed (see App. 1, Sec. G.2). The model statistics were ranked, and the four model structures with the highest-ranking statistics were selected for further analysis (Table A.6). Different numbers of the climate PCs, ranging from 2 to 20, were calibrated and verified for each type of structure. The calibration and verification statistics were tabulated, the significant results were selected, and the superior models were ranked to determine which ones were optimum.

There was often a general rise in calibration and verification statistics and then a general decline in verification statistics with increasing number of PCs, but very large fluctuations did occur from one PC number to the next and from one model structure to the next (see App. 1, Sec. G.2). To facilitate the selection procedure, the results for a given model structure were divided into three classes corresponding to models with 3 to 7 PCs, 8 to 13 PCs, and 14 to 20 PCs of climate. At least one model (the one with the "best" overall or highest-ranking statistics) from each class was examined closely and considered in the selection process. This practice assured that models with widely varying numbers of predictand PCs would be considered, but it did not guarantee that all three classes of models would be chosen. The final choice depended on the ranking of the model statistics.

2. Model Merging

Reconstructions from the highest-ranking models selected in the aforementioned fashion were averaged into combinations of two and three models of varying structure. Approximately twenty to thirty combinations were considered for each variable, season, and grid. The calibration and verification statistics were recalculated, and selected results were plotted for final model evaluation. At early stages in the work sometimes more than three models were averaged (Fritts, Lofgren, and Gordon 1979). Improvement in these larger models was so rare that it was decided to combine no more than three models for any one variable, grid, and season for the remaining part of the analysis.

Often there were two to four couplet or singlet models competing for the highest overall rank. Most combinations ranked higher than the individual couplets or singlets that entered into the average. Combinations with the greatest number of predictors in year t often had the highest-ranking statistics. The next most common predictors were for year $t + 1$, then $t - 1$, and then $t + 2$. Combinations that violated the biological or physical causal relationship, described in Fritts (1976) and in Fritts, Lofgren, and Gordon (1979), often but not always resulted in the worst calibration statistics.

3. Developing Models for Sea-Level Pressure

Several procedures had to be altered to develop, test, and select the optimum models for sea-level pressure. Subsample calibration and replication of sea-level pressure provided a complete set of independent data comparable to that of the dependent data. This information provided a continuous time series of independent data PCs, which were used for verification whenever possible, to avoid potential problems with spatial redundancy in the analysis of the gridded sea-level pressure data (Livezey and Chen 1983). The eigenvectors and their PCs were extracted from the covariance rather than the correlation matrix because the covariance seemed to portray a more meaningful representation of the associated patterns. The remaining procedures of calibration, verification, and model selection were the same as for temperature and precipitation (see App. 1, Sec. G.3, for more details on the sea-level pressure analysis).

6

The Statistics
of the Final Results

A. Characteristics of
the Selected Reconstructions

The statistics from the merged reconstructions and their components are summarized for seasons (Tables 6.1–6.4) and for annual values (Table 6.5). The N^* statistic, which is the df remaining in the residuals (out of a total of 61 to 63 df corresponding to the years used in the calibration, Eq. A.17), ranged from 28 to 57. Most models had 40 to 50 df remaining after calibration and merging. This number appeared to be adequate for the significance testing of annual values, but it was reduced to less than 10 when low-frequency features that exceeded time scales of 8 or more years were tested (see Eq. A.16). Thus, only the very largest low-frequency features within the period of calibration were statistically significant.

Generally, the statistics for the merged series were better than the statistics of the component series, and the statistics of the annual series were better than those of the seasonal merged series, indicating that the merging, on the average, did bring about improvement. The 46-grid temperature reconstructions exhibited the highest calibrated variance, and the 96-grid precipitation reconstructions exhibited the lowest calibrated variance.

The correlation coefficients and the sign tests, before and after taking the first difference, helped to identify differences in reliability between the low and high frequencies. For example, the lower values for the first difference data for annual pressure suggested that the low-frequency variations were reconstructed more reliably than the high-frequency variations. This feature was confirmed by the presence of significant

Table 6.1. Winter: Calibration and Verification Statistics of All Merged Models and Their Components

Model	Grid	EV	EV'	EV' max	RE'	N*	Coefficient of Contingency	Correlation	Correlation, First Differences	Sign Test	Sign Test, First Differences	Product Mean	Total Tests
											Verification Statistics[b]		
		Calibration Statistics[a]								% Tests Significant			
Pressure													
Merged models	96	0.376	0.284	0.658	0.195	45		45	45	36	27	10	42
Components													
15P15B15M		0.390	0.249	0.615	0.113	45		40	33	20	27	03	40
11P10B10I		0.287	0.195	0.689	0.107	43		55	18	36	18	05	33
11P12I		0.169	0.124	0.646	0.081	46		36	27	18	18	05	29
Temperature													
Merged models	77	0.416	0.287	0.481	0.162	44	0.314	52	56	24	20	21	38
Components													
5T15I		0.223	0.140	0.480	0.059	55	0.291	43	46	22	28	17	33
16T15I15M		0.449	0.301	0.611	0.157	41	0.311	50	46	13	19	18	32
Merged models	46	0.468	0.364	0.513	0.264	35	0.434	54	58	04	12	24	33
Components													
16T15B15I		0.421	0.292	0.635	0.168	36	0.407	27	38	04	19	24	23
8T15MB15I		0.254	0.150	0.494	0.052	35	0.425	42	50	23	08	26	28
Precipitation													
Merged models	96	0.204	0.034	0.350	−0.131	51	0.261	21	16	15	12	16	16
Components													
5R15I		0.127	0.034	0.402	−0.057	56	0.243	22	24	09	12	19	16
6R15I15M		0.222	−0.026	0.543	−0.267	47	0.239	16	16	15	08	16	14
Merged models	52	0.428	0.325	0.504	0.226	49	0.400	39	46	32	27	20	35
Components													
18R15MB15M		0.429	0.272	0.648	0.121	41	0.415	37	37	24	24	20	28
19R15I		0.292	0.212	0.463	0.134	56	0.363	34	41	22	17	22	27

[a] EV = explained variance; EV' = adjusted explained variance; $EV'max$ = maximum EV' in the climatic grid; RE' = adjusted reduction of error (estimate); N^* = number of independent years in the calibration period after adjusting for regression and autocorrelation of the residuals (Eq. A.29).
[b] Items are blank wherever value was not computed.
[c] Because of varying sample sizes and because the pressure tests were counted for only the number of PCs common to all model components, the percentages in this column are not always equal to the average of the five preceding columns.

power and cross-power spectral peaks at low frequencies (see below, Sec. F).

Fifty percent of the verification tests were significant for both the 77- and 46-grid estimates of annual temperature. This figure was higher than the percentages obtained for the verification of any seasonal data. The RE statistics greater than zero ranged from 52 percent for the 96-grid annual precipitation to 96 percent for the 46-grid annual tem-

| | Verification Statistics[b] | | | | | |
| | | Pooled Reduction of Error (RE) | | | | |
% RE > 0	RE	RISK	BIAS	COVAR	RE NEG = 0	PCs or Regions with RE > 0
27	−0.153	−0.507	0.011	0.343		5, 6, 10
0	−1.101					None
18	−0.294					10, 11
27	−0.125					2, 5, 11
57	−0.090	−0.478	0.111	0.277	0.181	5, 7, 8, 9
46	−0.155				0.145	5, 7, 8, 9
35	−0.350				0.143	6, 8
58	−0.050	−0.475	0.091	0.334	0.191	1, 2, 5
23	−0.353				0.120	None
62	−0.041				0.252	1, 5
44	−0.016	−0.106	0.007	0.084	0.093	1, 2, 3, 5, 9
55	0.008				0.056	1, 2, 3, 5, 7, 9
31	−0.131				0.116	1, 5, 9
59	0.059	−0.198	0.052	0.206	0.155	1, 2, 5
41	−0.023				0.129	1, 2, 5
56	0.009				0.125	1, 2

perature. These statistics were well above the averages for the seasonal data, but in some cases, such as those for the 77-grid autumn temperatures and the 52-grid summer precipitation, these statistics were higher for seasonal than for annual data. With the exception of annual precipitation, the pooled RE values appeared to be above that expected for random variation (Gordon and LeDuc 1981). Unlike the seasonal pressure reconstructions, the first 4 PCs of annual pressure all

Table 6.2. Spring: Selected Calibration and Verification Statistics of All Merged Models and Their Components

| | | | | | | | | Verification Statistics[b] | | | | | |
| | | Calibration Statistics[a] | | | | | | | % Tests Significant | | | | |
Model	Grid	EV	EV'	EV' max	RE'	N*	Coefficient of Contingency	Correlation	Correlation, First Differences	Sign Test	Sign Test, First Differences	Product Mean	Total Tests[c]
Pressure													
Merged models	96	0.402	0.286	0.611	0.175	51		50	10	50	10	08	32
Components													
15P15B15I		0.429	0.239	0.694	0.054	48		27	13	27	00	03	22
15P15I		0.261	0.170	0.578	0.082	57		40	07	33	13	03	36
10P10I10F		0.216	0.133	0.610	0.053	46		30	20	10	10	03	20
Temperature													
Merged models	77	0.437	0.335	0.625	0.238	40	0.308	23	45	15	28	19	28
Components													
16T15B15M		0.372	0.259	0.622	0.151	38	0.298	25	34	17	21	20	25
16T15I		0.338	0.272	0.620	0.208	44	0.296	30	38	13	21	19	23
16T15I15M		0.414	0.256	0.721	0.104	40	0.272	13	36	13	26	11	21
Merged models	46	0.512	0.430	0.577	0.352	29	0.456	48	52	12	32	23	32
Components													
13T15I15M		0.350	0.267	0.622	0.189	29	0.450	24	52	16	20	21	26
15T15I15F		0.420	0.305	0.628	0.196	31	0.401	36	36	12	16	23	22
10T15FMF		0.430	0.330	0.608	0.235	28	0.438	36	52	20	16	20	27
Precipitation													
Merged models	96	0.265	0.098	0.421	−0.064	49	0.236	13	16	12	08	14	12
Components													
10R15MBI		0.177	0.037	0.396	0.037	50	0.214	11	18	12	11	10	11
9R15I15M		0.221	0.008	0.460	0.008	49	0.195	12	13	11	07	11	10
10R15M15F		0.232	0.044	0.569	0.044	49	0.215	15	15	09	11	09	11
Merged models	52	0.356	0.217	0.460	0.083	50	0.347	15	22	17	17	28	17
Components													
11R15I		0.241	0.147	0.489	0.056	55	0.311	15	22	15	10	20	15
7R15I15M		0.250	0.054	0.508	−0.136	47	0.318	15	20	12	12	24	14
8R15I15F		0.328	0.109	0.590	−0.103	45	0.336	20	22	12	15	16	15

[a]EV = explained variance; EV' = adjusted explained variance; EV' max = maximum EV' in the climatic grid; RE' = adjusted reduction of error (estimate); $N*$ = number of independent years in the calibration period after adjusting for regression and autocorrelation of the residuals (Eq. A.29).

[b]Items are blank wherever value was not computed.

[c]Because of varying sample sizes and because the pressure tests were counted for only the number of PCs common to all model components, the percentages in this column are not always equal to the average of the five preceding columns.

had RE values greater than zero. These were also above the 0.95 level expected from random variations. Region 3, the Great Basin, was the only regional estimate of the 77-grid temperature reconstructions that did not have a positive RE statistic. All RE values for regions were positive for the 46-grid temperature reconstructions. Fewer RE statistics were positive for regions in the reconstructions for precipitation.

						PCs or Regions
% RE > 0	RE	RISK	BIAS	COVAR	RE NEG = 0	with RE > 0
			Verification Statistics[b]			
		Pooled Reduction of Error (RE)				

% RE > 0	RE	RISK	BIAS	COVAR	RE NEG = 0	PCs or Regions with RE > 0
40	0.017	−0.471	0.009	0.479		1, 3, 4, 5
00	−0.878					None
20	−0.128					3, 4, 10
30	−0.055					1, 3, 4
40	−0.036	−0.274	0.092	0.146	0.150	4, 9, 10
32	−0.165				0.081	4
32	−0.072				0.190	4, 7, 8
17	−0.218				0.258	4, 9
52	0.060				0.241	1, 2, 4, 5, 6
40	−0.004	−0.325	0.088	0.297	0.197	4
32	−0.223				0.152	6
40	−0.006				0.291	4, 5, 6
28	−0.108	−0.167	0.019	0.040	0.093	1, 6
21	−0.115				0.107	1, 6
24	−0.289				0.093	6
16	−0.270				0.096	6
32	−0.145	−0.302	0.047	0.110	0.188	1, 6
29	−0.112				0.186	1
24	−0.314				0.166	1, 6
24	−0.318				0.170	6

B. Spatial Patterns in the Statistics

The spatial features in calibrated variance and verification statistics for sea-level pressure are shown in Figures 6.1 through 6.3. A maximum in calibrated variance, centered over the Canadian Arctic, west of the mean position of the Icelandic Low, can be noted in all four seasons

Table 6.3. Summer: Calibration and Verification Statistics of All Merged Models and Their Components

Model	Grid	EV	EV'	EV' max	RE'	N*	Coefficient of Contingency	Correlation	Correlation, First Differences	Sign Test	Sign Test, First Differences	Product Mean	Total Tests[c]
Pressure													
Merged models	96	0.279	0.192	0.548	−0.108	55		27	00	27	09	07	20
Components													
11P15I		0.266	0.178	0.582	0.094	55		55	27	45	18	10	36
11P15F		0.234	0.141	0.629	0.051	56		18	00	27	00	03	13
Temperature													
Merged models	77	0.421	0.294	0.587	0.172	45	0.273	25	20	27	11	09	19
Components													
7T15I15F		0.403	0.170	0.682	−0.056	44	0.288	18	11	29	04	09	15
9T15F		0.332	0.226	0.573	0.124	51	0.246	15	13	16	07	11	13
8T15FMF		0.278	0.176	0.544	0.079	41	0.257	13	20	24	18	08	17
Merged models	46	0.451	0.345	0.605	0.243	45	0.358	22	26	22	11	17	20
Components													
16T15I		0.359	0.283	0.638	0.210	46		07	19	22	07	23	13
3T15F		0.330	0.112	0.592	−0.100	46		15	11	22	07	06	11
3T15FMF		0.380	0.078	0.694	−0.215	41	0.409	15	15	15	07	19	15
Precipitation													
Merged models	96	0.144	0.048	0.268	−0.045	55	0.204	16	18	15	06	12	13
Components													
16R15F		0.129	0.057	0.359	−0.014	56		18	12	11	07	09	11
6R15FFF		0.097	−0.083	0.313	−0.258	50		20	20	11	11	11	14
Merged models	52	0.075	−0.020	0.216	−0.112	55	0.295	20	24	22	05	24	17
Components													
4R15F		0.060	−0.001	0.337	−0.061	57	0.300	15	15	24	02	28	13
4R15FFF		0.082	−0.049	0.295	−0.176	53	0.323	29	32	27	10	28	22

[a]EV = explained variance; EV' = adjusted explained variance; $EV'\,max$ = maximum EV' in the climatic grid; RE' = adjusted reduction of error (estimate); $N*$ = number of independent years in the calibration period after adjusting for regression and autocorrelation of the residuals (Eq. A.29).

[b]Items are blank wherever value was not computed.

[c]Because of varying sample sizes and because the pressure tests were counted for only the number of PCs common to all model components, the percentages in this column are not always equal to the average of the five preceding columns.

and in the annual reconstructions. This area of maximum calibrated variance was verified in all but the spring season. Substantial pressure variance was both calibrated and verified for the southwestern United States; whereas little variance was calibrated and verified along the United States and Canadian border, where winter storms are most frequent (Bryson and Hare 1974). Large areas over the North Pacific Ocean were both calibrated and verified, except for the summer season, when smaller-scale patterns were evident.

| | | | Verification Statistics[b] | | | |
| | | | Pooled Reduction of Error (RE) | | | |
% RE > 0	RE	RISK	BIAS	COVAR	RE NEG = 0	PCs or Regions with RE > 0
27	−0.022	−0.360	−0.012	−0.326		1, 2, 7
27	−0.170					5, 6, 7
09	−0.252					1
49	−0.011	−0.129	0.053	0.065	0.180	1, 5, 6, 7, 8, 11
35	−0.094				0.218	1, 5, 6
42	−0.087				0.154	5, 6, 7, 8
44	−0.027				0.072	1, 5, 6, 11
63	0.072	−0.125	0.101	0.095	0.161	1, 3, 5, 6
37	−0.103				0.225	1, 5, 6
41	−0.054				0.089	6
44	−0.032				0.162	1, 5, 6
45	−0.031	−0.078	0.013	0.034	0.088	1, 3, 5, 9
36	−0.100				0.093	1, 3, 5
44	−0.046				0.077	1
66	0.032	−0.051	0.009	0.074	0.062	1, 2, 3, 4, 5, 6
56	0.000				0.063	1, 6
61	0.024				0.083	1, 3, 4, 5, 6

The sea-level pressure over some areas of eastern Asia appeared to calibrate and even to verify. Gordon et al. (1985) demonstrated a linkage between large-scale variations in reconstructed sea-level pressure over the North Pacific and winter severity in Japan. Lough et al. (1987) found sea-level pressure anomalies over the North Pacific that appeared to be correlated with the occurrence of widespread drought or floods in both Asia and North America. The lowest verification of the annual pressure series, however, can be noted in central Asia

Table 6.4. Autumn: Calibration and Verification Statistics of All Merged Models and Their Components

		Calibration Statistics[a]					Verification Statistics[b] % Tests Significant						
Model	Grid	EV	EV'	EV' max	RE'	N*	Coefficient of Contingency	Correlation	Correlation, First Differences	Sign Test	Sign Test, First Differences	Product Mean	Total Tests[c]
Pressure													
Merged models	96	0.362	0.261	0.582	0.164	47		42	25	33	08	10	36
Components													
12P12B12M		0.332	0.174	0.607	0.022	43		42	25	33	17	08	30
15P15I		0.277	0.201	0.648	0.127	51		40	13	27	07	07	34
Temperature													
Merged models	77	0.176	0.139	0.283	0.104	36	0.292	35	41	19	17	17	26
Components													
18T15I15M		0.062	0.035	0.179	0.010	38	0.222	24	07	07	00	13	09
17T15FMF		0.140	0.086	0.334	0.034	35	0.311	31	43	19	15	20	28
Merged models	46	0.328	0.252	0.523	0.179	35		19	11	07	07	27	10
Components													
13T15I15M		0.096	0.060	0.298	0.026	35		00	04	15	00	29	05
4T15FMF		0.351	0.134	0.697	−0.075	37		26	19	07	11	17	13
Precipitation													
Merged models	96	0.143	−0.146	0.290	−0.427	44	0.204	22	14	12	07	20	14
Components													
3R15IMF		0.142	−0.287	0.468	−0.702	40	0.222	24	16	14	05	17	15
3R15FMF		0.088	−0.095	0.429	−0.271	47	0.216	13	11	04	05	16	09
Merged models	52	0.227	0.036	0.308	−0.149	49		27	12	29	10	26	19
Components													
5R15I15M		0.161	−0.044	0.424	−0.243	49		10	17	15	07	22	12
5R15FMF		0.190	−0.012	0.394	−0.208	48	0.324	22	17	24	07	28	16

[a]EV = explained variance; EV' = adjusted explained variance; EV' max = maximum EV' in the climatic grid; RE' = adjusted reduction of error (estimate); N^* = number of independent years in the calibration period after adjusting for regression and autocorrelation of the residuals (Eq. A.29).

[b]Items are blank wherever value was not computed.

[c]Because of varying sample sizes and because the pressure tests were counted for only the number of PCs common to all model components, the percentages in this column are not always equal to the average of the five preceding columns.

through Siberia. The area of low verification extended east, then turns southeast over southern Alaska, the Gulf of Alaska, and the Pacific Northwest. This is an area of large pressure variations associated with the succession of cyclones and anticyclones that are carried in the upper-level flow near the zone of confluence (Lamb 1972). The cause of poor calibration in this area is not readily apparent.

Except for autumn, the calibration with temperature exceeded 30 percent over most of the grid (Figs. 6.4–6.8). The reconstructions of temperature were verified for many areas in the Plains states, western

			Verification Statistics[b]			
			Pooled Reduction of Error (RE)			
% RE > 0	RE	RISK	BIAS	COVAR	RE NEG = 0	PCs or Regions with RE > 0
33	−0.084	−0.415	0.004	0.327		1, 3, 4, 5
00	−0.437					None
20	−0.105					1, 4, 5
91	0.051	−0.031	0.022	0.060	0.062	1, 2, 3, 5, 6, 7 8, 9, 10, 11
44	−0.011				0.062	1, 2, 5, 6
72	0.039				0.085	1, 3, 6, 8, 9, 10
41	−0.045	−0.104	0.034	0.025	0.092	2, 6
59	0.003				0.054	1, 2, 6
26	−0.220				0.132	6
48	−0.063	−0.141	0.017	0.061	0.076	3, 4, 5, 6, 7, 8
46	−0.081				0.104	3, 4, 5, 7
39	−0.099				0.066	2, 3, 5
46	−0.039	−0.164	0.035	0.090	0.124	3, 4, 6
44	−0.096				0.095	6
34	−0.157				0.130	None

Great Lakes, and upper Mississippi Valley, but verification was poor in some areas of the Southwest and central western states. The annual temperature data verified the best, but three areas of poor reconstruction were indicated in the Intermontane Basin of the West, the Ohio River Valley to the Atlantic Coast, and an area stretching from southeastern Arizona through the Gulf states to the Atlantic Coast.

Good precipitation reconstruction for seasons was largely confined to the western states and to winter and spring. Calibration was poor in summer and autumn, but verification indicated acceptable reconstruc-

Table 6.5. Annual Calibration and Verification Statistics of All Merged Models and Their Components

Model	Grid	EV	EV'	EV' max	RE'	N*	Corre-lation	Corre-lation, First Differ-ences	Sign Test	Sign Test, First Differ-ences	Product Mean	Total Tests[c]
								Verification Statistics[b]				
		Calibration Statistics[a]						% Tests Significant				
Pressure[d]	96	0.484	0.402	0.723	0.322	49	60	30	50	20	60	44
Temperature	77	0.477	0.392	0.533	0.310	41	50					
	46	0.507	0.422	0.565	0.341	35	68	72	36	36	40	50
Precipitation	96	0.247	0.101	0.363	−0.040	51	27	23	18	07	18	19
	52	0.352	0.228	0.506	0.109	50	40	45	23	13	33	31

[a]EV = explained variance; EV' = adjusted explained variance; EV' max = maximum EV' in the climatic grid; RE' = adjusted reduction of error (estimate); $N*$ = average number of independent years in the calibration period after adjusting for regression and autocorrelation of the residuals (Eq. A.29).
[b]Items are blank wherever value was not computed.
[c]Because of varying sample sizes and because the pressure tests were counted for only the number of PCs common to all model components, the percentages in this column are not always equal to the average of the five preceding columns.
[d]Verification results were combined for only 10 PCs, the minimum number in common to all seasons.

tion in the Far West, central Plains, and Great Lakes during these seasons. Over 30 percent variance in annual precipitation was reconstructed in California, the Rocky Mountains, and the central Plains. The verifications did indicate that annual precipitation estimates were better than chance for some areas of the West. The area of reliable annual precipitation reconstructions was more dependent on the proximity of the tree-ring grid than was annual temperature and sea-level pressure. Although there appeared to be some reliability in the precipitation reconstructions for some stations to the east of the tree-ring grid, the statistics for annual precipitation indicated a general lack of reliability for stations throughout the South, eastern parts of the United States, and from eastern North Dakota to the northern Great Lakes.

The discrepancy between calibration and verification statistics may not be due entirely to decreasing reliability of the reconstructions with independent data. It could also arise in part from the poor quality of the independent instrumental measurements (described in Chap. 3), urbanization, pollution, or nonclimatic factors affecting the forest environments (Fritts and Swetnam 1989).

Verification Statistics[b]						
Pooled Reduction of Error (*RE*)						
% *RE* > 0	*RE*	RISK	BIAS	COVAR	*RE* NEG = 0	PCs or Regions with *RE* > 0
60	0.167	−0.342	0.008	0.501		1, 2, 3, 4, 6, 7
75	0.127	−0.254	0.210	0.171		1, 2, 4, 5, 6, 7 8, 9, 10, 11
96	0.225	−0.276	0.174	0.327		1, 2, 3, 4, 5, 6
52	−0.001	−0.113	0.031	0.081		1, 3, 5, 6, 7, 9
60	0.005	−0.214	0.012	0.207		1, 2, 5, 6

C. Sources of Variation in Calibration Statistics

It is important to remember that the reconstructions used in the following analysis were estimates of seasonal temperature, precipitation, and sea-level pressure made at points or stations over a spatial grid for the years from 1601 to 1963. The methods used seasonal and annual rather than monthly climatic data, emphasized the large-scale patterns by using the largest eigenvectors and their PCs, and obtained regressions using only the largest and most significant canonical variates. Although the reconstructions were estimated at individual grid points, the information content was biased toward large-scale features. A different calibration strategy would be required to maximize the information at the individual grid points.

To evaluate these large-scale features, the data were stratified, pooled, and averaged in several ways and the statistics recalculated (Fig. 6.9). The statistics of the various averaged sets were compared to each other and to the statistics for the individual components over the grid points. The stratifications produced six categories of averages in addition to the

VERIFICATION STATISTICS

CALIBRATED VARIANCE

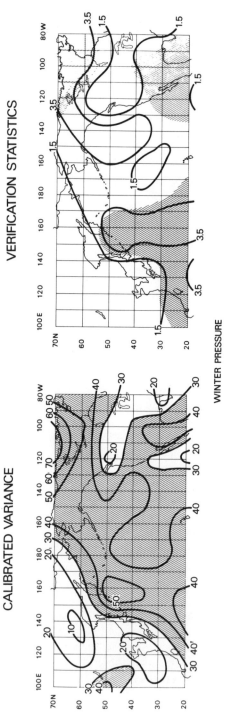

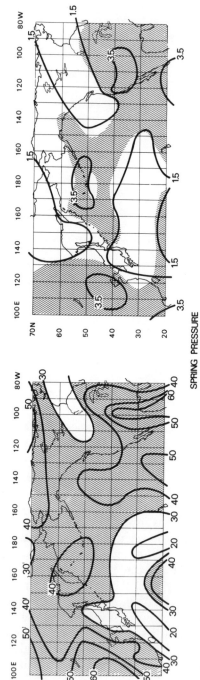

WINTER PRESSURE

SPRING PRESSURE

Fig. 6.1. The percentage of calibrated variance and the number of verification statistics out of a total of 5 for the merged models used to reconstruct the 96-grid winter and spring sea-level pressure. *Shading*: grid points with calibrated variance of 30 percent or above (*left*) or positive *RE* (*right*).

VERIFICATION STATISTICS

CALIBRATED VARIANCE

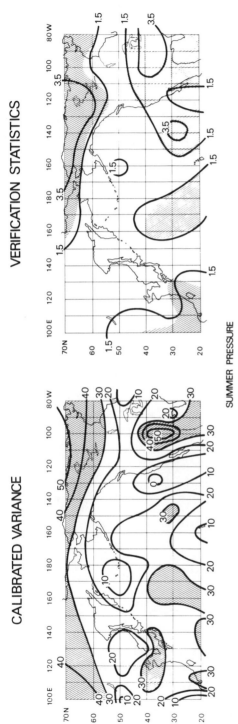

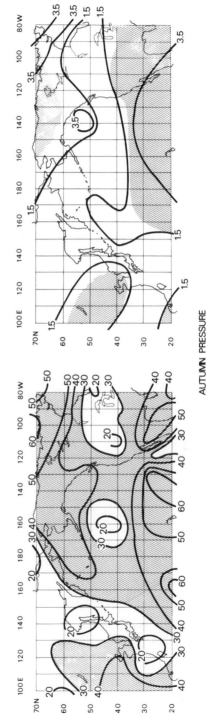

SUMMER PRESSURE

AUTUMN PRESSURE

Fig. 6.2. Same as Figure 6.1, for summer and autumn sea-level pressure.

CALIBRATED VARIANCE

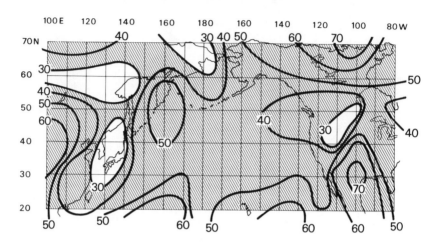

VERIFICATION STATISTICS

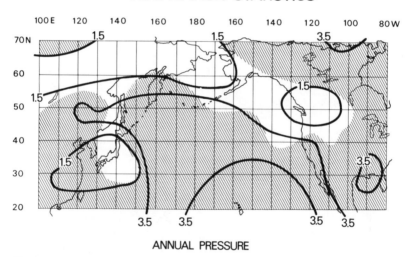

ANNUAL PRESSURE

Fig. 6.3. Same as Figure 6.1, for annual sea-level pressure.

individual components shown at the bottom of Figure 6.9. These categories included (1) pooling the results from each model (individual components) to obtain an average merged model (Tables 6.1–6.4) while preserving the grid-point statistics (Figs. 6.1, 6.2, and 6.4–6.7), and (2) pooling the merged models for each season to obtain the annual results (Table 6.5) while preserving the grid-point statistics (Figs. 6.3 and 6.8).

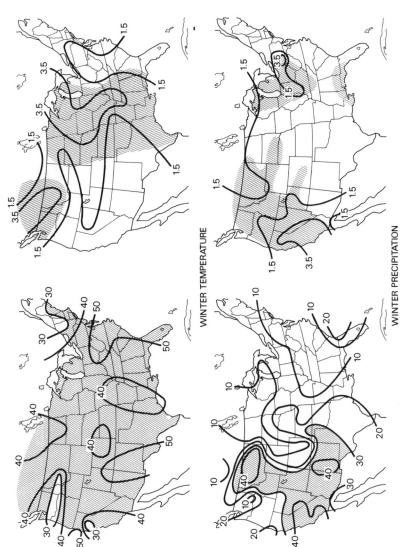

Fig. 6.4. The percentage of calibrated variance, the number of significant verification statistics out of a total of 5, and presence of a positive *RE* statistic for the merged models used to reconstruct 77-grid winter temperature (see Fig. A.1 and A.2) and 96-grid winter precipitation.

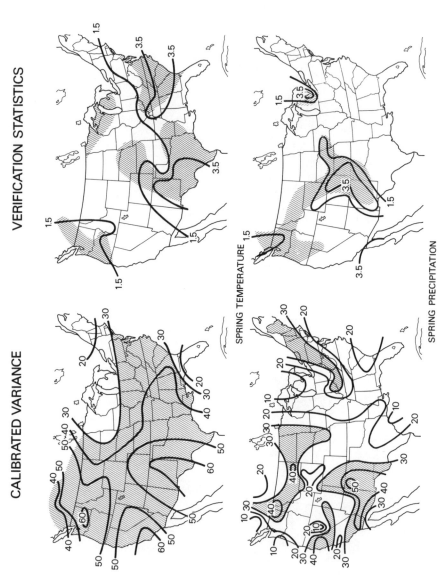

Fig. 6.5. Same as Figure 6.1, for 77-grid spring temperature and 96-grid spring precipitation.

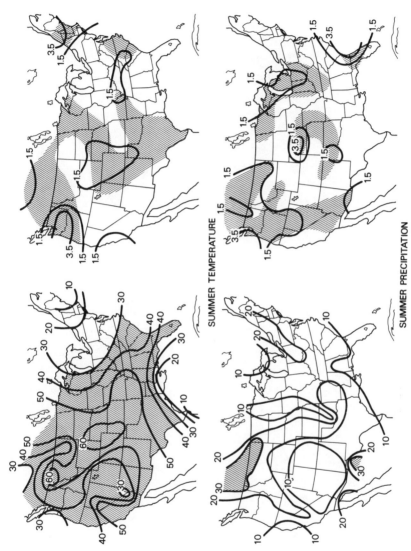

Fig. 6.6. Same as Figure 6.1, for 77-grid summer temperature and 96-grid summer precipitation.

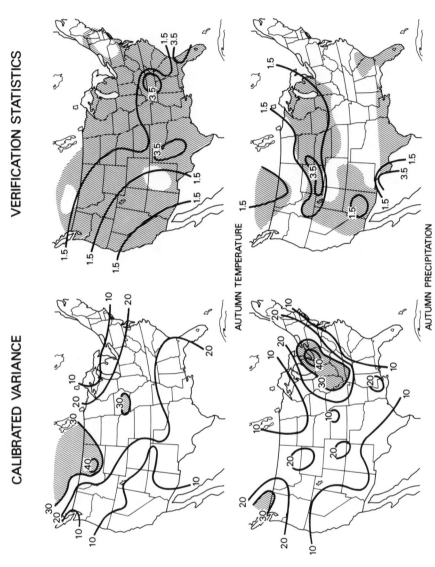

Fig. 6.7. Same as Figure 6.1, for 77-grid autumn temperature and 96-grid autumn precipitation.

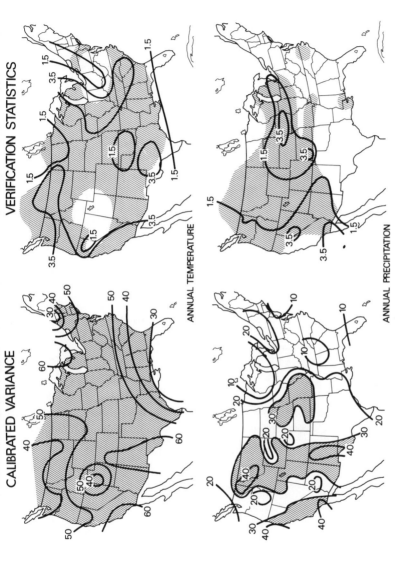

Fig. 6.8. Same as Figure 6.1, for 77-grid annual temperature and 96-grid annual precipitation.

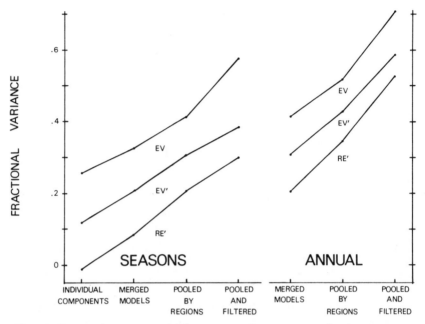

Fig. 6.9. The explained variance (*EV*), adjusted explained variance (*EV'*), and estimated *RE'* averaged over all variables, seasons, and grids, and plotted from left to right for (a) individual seasonal component models, (b) merged seasonal models, (c) merged seasonal models pooled and averaged by region, (d) seasonal pooled averages filtered, (e) annual merged models, (f) annual merged models pooled and averaged by region, and (g) annual pooled averages filtered.

In addition, the results from stratification 1 and 2 were stratified further by (3) pooling and averaging the grid-point values within each region (Figs. 3.1 and 3.2, Table 3.1) and (4) treating the regional averages with a low-pass filter (see Figs. 7.2–7.4) to bring out the low-frequency features.

1. Pooling All Variables and Grids

Three calibration statistics, *EV*, *EV'*, and *RE'*; the five verification statistics that could be tested for significance; and the *RE* were considered in the following analysis (Table 6.6). As expected, the adjusted fractional variances (*EV'*) were always smaller than the unadjusted fractional variance (*EV*) values, because the former "is the expected value of *EV* in a (hypothetical) large population from which the dependent data sample was taken" (Kutzbach and Guetter 1980 p. 176) (see Eq. A.16). The term *RE'* is an estimate, called the *adjusted reduction in error*, calculated from the dependent data; "in the absence of *RE* statistics, the value of *RE'* is an indication of whether a regression model will yield good predictions on independent data" (Kutzbach and Guetter 1980 p. 176) (see Eq. A.27). The *RE'* is always smaller than the *EV* and *EV'* statistics.

The averages of *EV*, *EV'*, and *RE'* in each column of Table 6.6 increase significantly ($p = 0.95$) from top to bottom, with increasing numbers of data included in the averaging. They also increase for the three pooling categories from left to right. These data are rearranged and plotted in Figure 6.9 to show the change in fractional variance representing improvement.

The steeper the slope in the figure, the greater the effect of that particular source on the calibrated variance. Pooling of the annual *EV'* values led to the greatest increase in variance, and the merging of the individual components for seasons led to the least increase in variance. The means of statistics for regional data, however, before and after filtering, rarely proved to be significant. This result could be due to limited sample size and too few *df* for adequate testing, or it could mean that the filtering did not contribute to improvement. For this reason, it was not certain from these statistics that filtering improved the estimates. Nevertheless, the average values were larger for the

Table 6.6. Calibration Statistics for all Reconstructed Variables Stratified at Several Levels, Pooled, and Averaged

Models	Pooling Grid Points								
	Stations / Grid			Regions			Regions Filtered		
	EV	EV'	RE'	EV	EV'	RE'	EV	EV'	RE'
Component models	0.255	0.127	−0.006						
Merged for seasons	0.324	0.203	0.087	0.411	0.307	0.207	0.574	0.384	0.299
Annual combinations	0.413	0.309	0.208	0.515	0.429	0.346	0.703	0.585	0.524

Note: The statistics are averaged over all models, grids, seasons, and variables. The data for the component models are not averaged by region.

filtered data, and the changes in df were estimated conservatively (see Eq. A.28). This question was evaluated further by means of spectral analysis (see Chap. 6, Sec. F.2).

If we assume for the moment that filtering the annual values over regions produced a real improvement, the average EV' values (Table 6.6) show that 42.9 percent of the unfiltered variance was calibrated over all regions and that 58.5 percent of the filtered variance was calibrated. This average included all variables, grids, and data points, but it did not include the high-frequency variations. Adjusted EV values reported in other studies ranged from 38 to 60 percent (Briffa, Jones, et al. 1988; Briffa, Jones, and Schweingruber 1988; Cook and Jacoby 1977; Stahle and Cleaveland 1988; Stahle et al. 1988; Meko 1982; Meko et al. 1980; Michaelsen et al. 1987). The results from Table 6.6 compare well with the variance calibrated in the other dendroclimatic studies, even though the other studies optimized the calibration only for one particular region. Only the EV' data are shown in the remaining figures.

2. Differences Attributed to Climatic Variables

The aforementioned statistics also were compiled and averaged separately for precipitation, temperature, and sea-level pressure to assess the differences among the three climatic variables (Table 6.7, Fig. 6.10). The lower calibration for precipitation is apparent. The average explained variance for the individual components and the seasonal merged models for precipitation were the only ones that averaged less than 0.10. They were not much higher than noise-level estimates, and the average differences were not significant ($p \geq 0.95$). The improvement in fractional variance, however, when the reconstructions of precipitation were pooled by region, was 0.156 for seasons and 0.189 for annual totals. These figures indicated more improvement than was observed for either sea-level pressure or temperature; however, the improvement in EV' was not significantly different from zero ($p \geq 0.95$). Additional but insignificant improvement in the precipitation calibration was apparent when the regional data were pooled to obtain filtered values.

The fractional variance calibrated for temperature was higher than it was for precipitation. This finding suggests that temperature was at least as important to growth in arid-site trees as precipitation, if not more so, through the effect of temperature on evapotranspiration. High temperatures are often associated with low precipitation, though, and it may be impossible to separate the independent effects of these two variables. On the average, the spatial variations in temperature always

Table 6.7. Calibration Statistics for each Reconstructed Variable Stratified at Several Levels, Pooled, and Averaged

	Pooling Grid Points								
	Stations / Grid			Regions			Regions Filtered		
Models	EV	EV'	RE'	EV	EV'	RE'	EV	EV'	RE'
Pressure									
Component models	0.285	0.180	0.078						
Merged for seasons	0.355	0.256	0.161	0.406	0.314	0.226	0.712	0.618	0.570
Annual averages	0.484	0.402	0.322	0.525	0.449	0.376	0.777	0.706	0.667
Temperature									
Component models	0.309	0.197	0.078						
Merged for seasons	0.402	0.306	0.215	0.465	0.380	0.300	0.610	0.349	0.269
Annual averages	0.492	0.407	0.326	0.566	0.493	0.424	0.762	0.621	0.569
Precipitation									
Component models	0.187	0.021	−0.146						
Merged for seasons	0.231	0.074	−0.077	0.359	0.230	0.105	0.469	0.304	0.193
Annual totals	0.300	0.165	0.035	0.459	0.354	0.254	0.608	0.488	0.408

Note: The statistics are averaged over all models, grids, and seasons. The data for the component models are not averaged by region.

appeared to be reconstructed more accurately than the spatial variations in precipitation at all levels of pooling. With the exception of the filtered data, the improvements from averaging were more often statistically significant for temperature than for precipitation. Filtering appeared to reduce rather than to increase the fractional variance for seasonal temperature, and pooling of the annual data by region did not provide significant improvement for temperature.

The improvement in fractional variance of seasonal pressure between the individual components and merged models, and merged models and

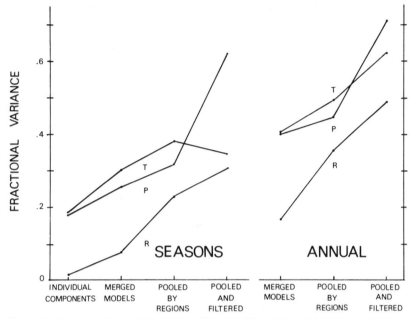

Fig. 6.10. Same as Figure 6.9, except with only *EV'* and the data separated into reconstructed temperature (*T*), precipitation (*R*), and sea-level pressure (*P*).

data pooled by region, was less than the improvement for temperature, and the average *EV'* values were smaller. The filtered values for fractional variance of seasonal sea-level pressure were much improved and higher than the filtered values for the other two variables, but the differences between the filtered and unfiltered regional averages were not large enough to be significant. The unfiltered temperature reconstructions appeared to be superior to the reconstructions of pressure, whereas the filtered regional pressure reconstructions appeared to be superior to the reconstructions of temperature. The same relationships hold for the annual data, and the improvement was significant ($p \geq$ 0.95). This result supports the finding that the low-frequency variations in reconstructed sea-level pressure appeared substantially more reliable than the high-frequency variations in reconstructed sea-level pressure.

3. Differences between Large and Small Grids

The data for temperature and precipitation were divided into small and large grids, the statistics were calculated, and the values were averaged to evaluate grid-size differences (Fig. 6.11). As might be expected, the

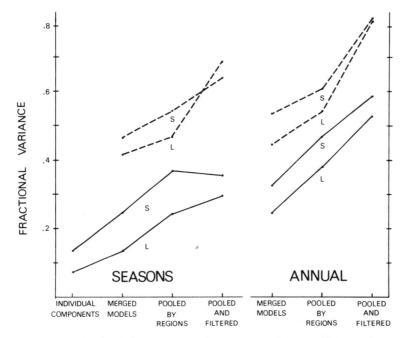

Fig. 6.11. Same as Figure 6.9, except that the *solid line* is *EV'*, the *dashed line* is the average maximum *EV'*, and the data are separated into large (*L*) and small (*S*) grids.

average EV' for the large grids, which included a large number of stations at great distances from the chronology sites, was lower than the average EV' of the small grids. The differences were on the order of 6 to 7 percent. Similar relationships can be observed in the EV' *max* figures, except that the differences between the large and small grids became small or reversed sign when the data were pooled into regions and these values filtered. It is clear from the data in Figure 6.11 that both the maximum and the average calibrations were higher for the small grids than for the large grids. As other evidence in Chapter 6, Section E, suggests, however, reconstructions for some stations outside the area of the small climatic grids were better than expected by chance.

4. Differences among the Seasons

The reconstructions for all variables and grids were pooled by season, the statistics were recalculated, and the EV' fractional variance was plotted (Fig. 6.12). The averaged annual values are included in the

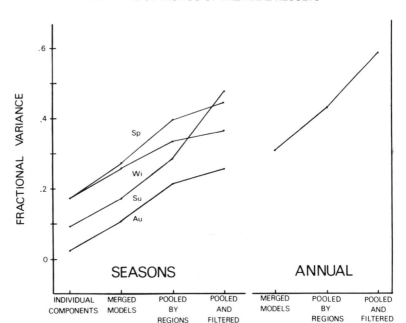

Fig. 6.12. Same as Figure 6.9, except with only *EV'* and the data separated into seasons. The annual averaged data shown in Figure 6.9 are included for comparison.

figure on the right to facilitate comparison. The differences in fractional variances (*EV'*) between seasons were highly significant ($p \geq 0.99$) when both the components and the merged models were tested. The differences were insignificant when only the merged models were tested, probably because too few observations (5) were present. The number of component models within each season and variable was too small for individual testing. The fractional variance calibrated for individual components was highest for both winter and spring. Those for summer and particularly autumn were much lower. The winter reconstructions did not improve as much with merging as the spring reconstructions, and the regional reconstructions were poorer. The *EV'* for summer improved at the same rate as spring when the models were merged and when the data were pooled by region. When the summer data were filtered, however, the fractional variance exceeded the variance of the filtered spring values. This result suggests that a greater proportion of the large-scale and low-frequency variations appears to be well calibrated in the summer season. The reconstructions for autumn were inferior to those for the other three seasons. The reconstructions for all seasons were substantially improved by combining them into annual values.

5. Differences among Regions

The average EV' statistics from the merged temperature and precipitation models were calculated for each region and grid (Fig. 6.13). There was always some increase on the plots between the average EV' values for the merged models and the average values for data pooled by region for both the seasonal and annual data. This observation suggests that the pooling of data throughout these regions improved calibration consistently. The variability in calibration for different regions was considerably greater after the data were filtered. The precipitation results varied more from region to region than the results for temperature (Fig. 6.13). The EV' for filtered annual precipitation reconstructions ranged from negative values for the eastern regions 9, 10, and 11 to values exceeding a fractional variance of 0.8 for region 3, the Great Basin. The precipitation for regions 1 and 2, along the Pacific Coast, was moderately well calibrated for the 96-grid but poorly calibrated for the 52-grid. Low-frequency variations in region 4, the southwestern states, were very poorly calibrated, especially in the 96-grid analysis. Regions 5 and 6, along the eastern slopes of the Rocky Mountains, were relatively well calibrated, as well as regions 7 and 8, which are in the central Plains east of the tree sites. The low-frequency variations in region 9 were surprisingly well calibrated considering the distances between these regions and the tree sites. Most evidence suggests that the reconstructions for the Appalachians and the Atlantic and Gulf coasts were not significantly different from random estimates.

Part of this high variability may be natural random variability of the low-frequency information in the time series, which became more apparent as the numbers of df were reduced by filtering. Other possible explanations for these differences include (1) changes in the low-frequency responses of some forest stands in a region related to stand history; (2) differences in precipitation or temperature trends in the mountains, where the trees usually were located, as opposed to the valleys, where the climatic stations were located; (3) features inherent in the atmospheric circulation for different seasons which may contribute to the variability; or (4) variable interactions among the chronology responses to the climate in different seasons. On the average, except for seasonal temperature, there was improvement in the fractional variance explained when the regional data were filtered.

Sea-level pressure had the highest EV' pooled by region (PCs) and filtered. In the case of pressure, the statistics were calculated from the time series of the first 10 PCs with highest variance. This procedure is probably a more efficient way to partition the data than to average values over geographic regions, and it is not surprising that the analysis

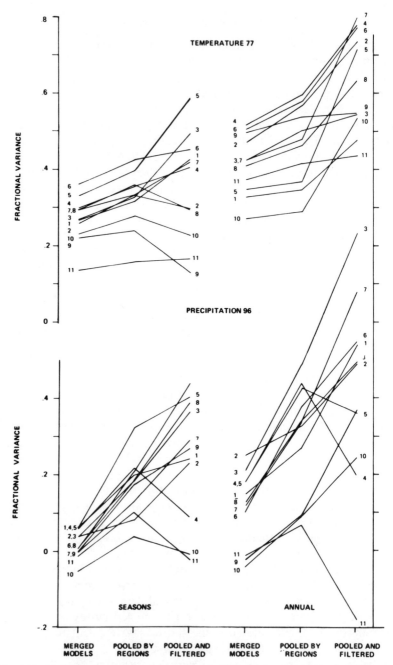

Fig. 6.13. Same as Figure 6.9, except with only *EV'* and the data separated by variable, grid, and region.

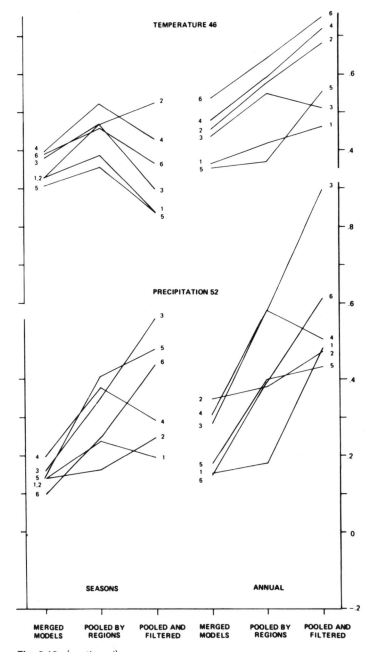

Fig. 6.13. (continued)

of PCs accounted for a higher percentage of variance than the regional averages of temperature and precipitation.

D. Changes in the Verification Statistics

1. Percentages of Significant Statistics

The percentages of verification statistics that could be tested for significance and the *RE* values greater than zero were classified, averaged, and compared in the same way as the calibration statistics, except that the independent instrumental record varied in length for each station of the temperature and precipitation grids. Sea-level pressure was the only variable with continuous and complete estimates for all grid points in

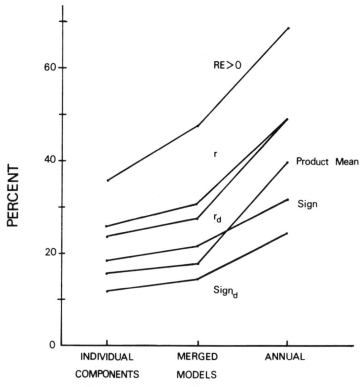

Fig. 6.14. The verification statistics of *RE* greater than zero, correlation coefficient (*r*), correlation of the first difference (*r*$_d$), product mean, sign, and sign of the first difference (*Sign*$_d$), stratified and averaged for individual components, merged models, and annual combinations.

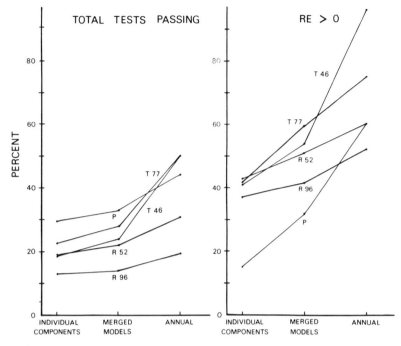

Fig. 6.15. Same as Figure 6.14, for the total number of verification tests that were significant and the number of *RE*s > 0 separated by variable and grid.

the independent period (1900–1963). Therefore, the regional averages (i.e., the PCs) were available only for sea-level pressure.

The average results for each verification test are shown in Figure 6.14. All but the sign test were highly significant ($p \geq 0.99$), and the sign test was significant at the 0.95 p level. Increasing numbers of all tests passed with each successive level of averaging. The *RE* statistics were greater than zero 36 to 68 percent of the time, but smaller numbers of the r and r_d tests were significant. Significant sign tests for individual components and merged models were the next most frequent. The product mean tests on the annual data were significant more often than the sign tests, but fewer product mean tests were significant for the individual components and merged models. Both sign tests and the product mean test appear to be the least discriminating of the statistics.

The low numbers of significant product mean tests for the seasonal analyses were consistent with the conclusions of Gordon and LeDuc (1981) that the test is conservative, especially when *EV* is low. There was a substantial improvement of the product mean tests over the sign tests for the annual reconstructions, however. In fact, the improvements in all four parametric tests were substantially greater than the improvements of the nonparametric sign tests.

Figure 6.15 summarizes the verification results for each climatic variable and grid. The differences between individual components and merged models were insignificant, but those between the merged and annual models were significant ($p \geq 0.95$) for the RE greater than zero and highly significant ($p \geq 0.99$) for the total number of tests passing. The improvement in statistics due to merging was substantially greater for temperature than for either precipitation or pressure, especially for the annual averages. A greater percentage of independent statistics was significant for the large temperature grid than for the small temperature grid, and the percentages of RE greater than zero were higher for the individual components and merged models. No differences were noted in the total tests passing for the two grids of annual temperature, and the smaller temperature grid had a greater percentage of RE greater than zero. This result indicated that the large temperature grid verified almost as well as the small temperature grid. The large precipitation grid was clearly inferior to the small precipitation grid. The

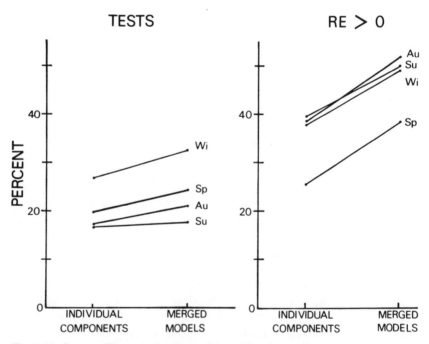

Fig. 6.16. Same as Figure 6.15, separated into different seasons.

sea-level pressure models for seasons had the highest numbers of total tests passing, but the *RE* statistics were very poor, suggesting that some seasonal reconstructions of sea-level pressure are not accurate (see App. 1, Sec. G.4). In all cases, the annual reconstructions were superior to the seasonal reconstructions.

The independent verification statistics that were significant and the *RE* greater than zero for the individual components and merged models for the four seasons (Fig. 6.16) indicated that more verification tests were significant for winter than for the other three seasons. There was essentially no improvement in the statistical verification tests for summer attributable to merging models. The percentage of *RE* greater than zero for spring was clearly lower than for the other seasons. These results suggest that the reconstructions for the different seasons have variable characteristics, though none of the four seasons can be rejected outright as unacceptable. Other information suggests that the reconstructions for autumn probably are the least reliable. All of the other seasons had some desirable characteristics. The differences between seasons need to be investigated more thoroughly.

2. Pooled-*RE* Statistics

The *RISK*, *BIAS*, and *COVAR* terms for the *RE* statistics were averaged and the results plotted for all variables and grids and then for each grid and variable (Fig. 6.17; also see Eq. A.21 and A.22). As shown in the previous figures, the values of *RE* generally rose as individual components were merged, and these data were combined into annual averages. The *COVAR* term, shown by the vertical size of the shaded area, is larger than the *BIAS* term, shown by the vertical size of the unshaded area. The average *RE*s were negative for individual components for all variables and grids, indicating that some reconstructions were very poor in every grid that was reconstructed. With the exception of the small temperature grid, the average *RE*s also were negative for the merged models. All average *RE*s for the annual reconstructions were positive except those for the large precipitation grid. These data also support the conclusion that the reconstructions improved as more and more data were averaged and that the annual estimates for the most part were superior to the seasonal estimates.

The *BIAS* term was large for the annual values of the large temperature grid. This finding suggests that the differences in the annual temperature trends from the twentieth-century average values were better verified than the year-to-year variations in the 77-grid reconstructions. Some low-frequency trends were also verified in the 46-grid, as indicated by the large *BIAS* term, and the year-to-year variations

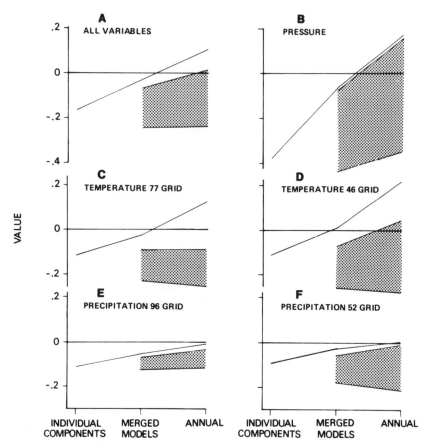

Fig. 6.17. *RE* statistic (*line*), *RISK* term (*lower boundary of shaded area*) and sum of the *RISK* and *COVAR* terms for all variables (*A*) and for individual variables and grids (*B*)–(*F*) (*upper boundary of shaded area*). The *shaded area* is proportional to the *COVAR* term, and the *unshaded area* above it is proportional to the *BIAS* term. Only the *RE* statistic was calculated for individual components.

measured by the *COVAR* term were reconstructed accurately. Differences in the length of the independent climatic records from the West compared to those in the East may have contributed to some of the verification differences. This result suggests that the low-frequency annual temperature reconstructions for eastern North America may have been reconstructed more accurately than the high frequencies. Both the 46- and 52-station climatic grids that were, on the average, closer to the sites of the trees had higher average *RE* statistics than the

77- and 96-station grids. Again, the temperature and sea-level pressure grids verified better than the precipitation grids.

The low amounts of variance associated with the *BIAS* term for sea-level pressure suggests that the verification procedures for sea-level pressure may have underestimated the errors in the low frequencies. For example, the correction made for the mean differences between the dependent and independent estimates (Chap. 5, Sec. G) may have eliminated some of the low-frequency variability. The spectral analysis and other evidence indicated that some reliable low-frequency information appeared to be present in the reconstructions of sea-level pressure even though it was not part of the *BIAS* statistic. The fact that the *BIAS* terms were lower for precipitation than for temperature suggests that (1) long-term variations were less evident for precipitation than for temperature, (2) tree-ring chronologies did not retain a reliable record of long-term trends in precipitation, or (3) the long-term temperature changes had a greater effect on growth than precipitation changes and may have masked the effects of long-term precipitation changes.

The *RE* statistics for temperature and precipitation were averaged and pooled by region for each variable and grid (data not included here) to examine the regional variations. Regions in the western half of the United States and Canada (regions 1–6), which were generally closer to the tree sites, had higher *RE* and *COVAR* statistics than the eastern regions. These findings provided additional evidence that a higher amount of the variance was reconstructed and verified in the western than in the eastern regions. The *BIAS* for temperature was largest in regions 5 and 6, east of the Rocky Mountain crest, and the *BIAS* contributed substantially to positive *RE* statistics for regions 7, 8, and 9, the northern and southern Plains and the western Great Lakes. This result confirms that some temperature trends were reconstructed and verified for the five areas to the east of the Rocky Mountains.

3. *RE* Statistics for PCs of Pressure

The *RE* and its three components were calculated for each of the PCs of sea-level pressure before and after the PCs were treated with a low-pass filter. The results for the *RE* statistic and the three components were plotted for the first 5 PCs in Figure 6.18, using the same representation as in the preceding figures. The *RE* values for the first 4 PCs of annual pressure were positive. The *RE* for PC 5 became positive after the high-frequency variations were removed by filtering. Unlike the results for temperature and precipitation, sea-level pressure in winter was not reconstructed as accurately as in other seasons. Only the

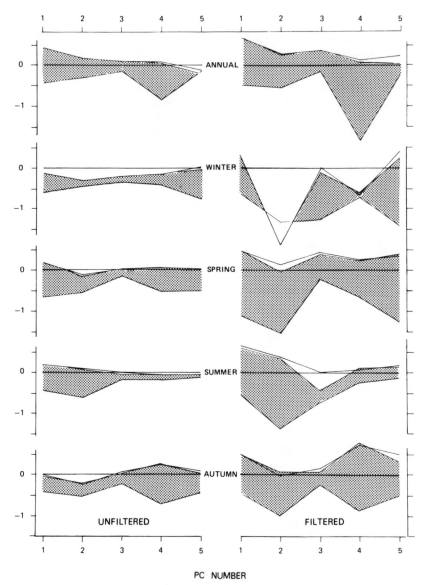

PC NUMBER

Fig. 6.18. Same as Figure 6.17, except that the unfiltered and filtered annual and seasonal reconstructions of the first 5 PCs of sea-level pressure were analyzed.

filtered PCs 1, 3, and 5 could be verified with the *RE* statistic, and the *RE* for PC 3 was barely positive. All but PC 2 for spring were verified with positive *RE* values. The reconstructions for all three variables were relatively reliable for the four-month spring period. Both the filtered and unfiltered PCs 1 and 2 for summer were verified, and the filtered PCs 4 and 5 had positive *RE* values. All autumn PCs, except the unfiltered PC 2, had positive *RE* statistics. The *RE* statistics were more often positive and large for the filtered than for the unfiltered pressure reconstructions, and the individual components varied more widely. The higher *RE* values for the filtered data support the conclusion that the low-frequency and large-scale variations in pressure were reconstructed most accurately.

E. Regional Differences between Grid Sizes

The averages of the calibration and verification statistics for the small grids were generally superior to those for the large grids. Direct comparisons had not been made for reconstructions from the same general areas of western North America, however. To examine these differences more carefully, the grid-point annual temperature and precipitation statistics were averaged for regions 1–6 (Table 6.8). In addition, the averages of statistics for regions 7–9 in the large grid were calculated so they could be compared to the data from regions 1–6. Regions 10 and 11 were not considered, as they had no reliability.

The fractional variance for annual temperature in regions 1–6, after adjusting for *df* loss (EV'), was 0.422 for the small grid, 0.392 for the large grid, and 0.415 for regions 7–9 to the east of the tree-ring chronologies. These figures are remarkably similar. For the 46-temperature grid, 50.4 percent of the verification statistics were significant in regions 1–6, compared to 54.6 percent in regions 1–6 of the 77-temperature grid, but the average *RE* statistics were higher. The percentages of the sign and the product mean tests for temperature were higher for regions 7–9 than for the 46-grid temperature, and the average pooled + *RE* statistics were higher. Both temperature grids had strengths and weaknesses in the West. Only the large grid had information for regions 7–9, and the reconstructions for these regions appeared to be of similar or slightly lower quality to those of regions 1–6.

The individual verification statistics for temperature in regions 1–6 suggest that the two grid sizes produced some interesting differences in the temperature reconstructions for the West. More correlation coefficients and more sign tests were significant for the 46-grid than for the

Table 6.8. Grid-point Calibration and Verification Statistics for Annual Temperature and Precipitation Averaged for Particular Regions

	Temperature			Precipitation		
	Grid			Grid		
	46	77 Regions	77	52	96 Regions	96
Tests	1–6	1–6	7–9	1–6	1–6	7–9
Calibration tests						
EV	0.507	0.441	0.514	0.352	0.315	0.220
EV′	0.422	0.392	0.415	0.228	0.181	0.066
Verification tests (%)						
Correlation (r)	68.0	54.4	42.2	40.0	35.1	34.7
r of first difference	72.0	86.4	58.0	45.0	30.1	26.5
Sign test	36.0	27.2	41.8	22.5	25.1	19.2
Sign of first difference	36.0	68.2	26.4	12.5	10.0	7.8
Product means	40.0	36.4	52.6	32.5	25.2	19.2
All tests	50.4	54.6	44.2	30.5	25.0	21.4
Average RE	0.254	0.146	0.221	0.007	0.061	0.005
Pooled RE	0.225	0.167	0.193	0.005	0.051	0.022
Pooled + RE	0.252	0.212	0.290	0.187	0.166	0.110

77-grid, but more correlations of the first difference and more signs of the first difference were significant for the 77-grid. This result suggests that more of the low-frequency variations in climate were verified in the West using the 46-grid, whereas more high-frequency variations were verified using the 77-grid. The low-frequency verification in the 77-grid noted in Figure 6.17.C must have come from stations in the East.

All but one of the statistics for regions 1–6 from the 52-grid precipitation reconstructions were superior to those for regions 1–6 from the 96-grid. The statistics for precipitation in regions 7–9 were generally lower. These results indicate that the precipitation reconstructions outside of the area of the tree-ring grid were substantially less accurate than the corresponding reconstructions of temperature. At best, the reconstructions of only a few precipitation stations in regions 7–9 were acceptable.

The relationships associated with merging reported for the calibration statistics can also be observed in the verification statistics shown in Table 6.8. The annual values of temperature appeared to be the most reliable. Though the statistics from the smaller grids frequently had the highest average values, the statistics from the larger grids, particularly those for temperature, indicated that many large grid estimates were reliable, especially in the area of the Great Plains. Thus we chose to present the results from the large grids with the caveat that all grid points east of the Great Lakes and Mississippi River were unreliable, many temperature estimates in regions 7–9 were reliable, and precipitation estimates for only a few stations in the Great Plains were reliable.

F. Variance Analysis

The preceding chapter described the reconstructions as a set of mean patterns or filtered values plotted over space and time. It is also meaningful to consider the variances as well as the mean values. How do the variances change when the reconstructions from several models are averaged, grid points are combined into regional averages or filtered, and seasonal data are combined to obtain annual values? The variances and spectral densities were calculated for the various combinations of the reconstructions to help answer this question.

1. The Effects of Averaging on the Variance

The square roots of the variances (the standard deviation) for the 1901–1961 instrumental data and the reconstructions were used to express variation in terms of the original temperature, precipitation, and sea-level pressure measurements (Figs. 6.19 and 6.20). As one would expect from the Central Limit Theorem, the variances and standard deviations declined for each successive level of pooling. The decline in variance is a function of the level of averaging and is usually associated with an increasing percentage of explained variance of the estimates (Fig. 6.9). The instrumental data were the same for the individual components and merged models (Figs. 6.19 and 6.20), so there was no difference between the variances of the instrumental data for the two classes. The differences in variance between the two grid sizes were greater for precipitation than for temperature, because the precipitation units were expressed as percentage of the average value of the calibration period, and the average variance of precipitation percentage in the dry West was much higher than the variance of precipitation percentage in more moist eastern regions.

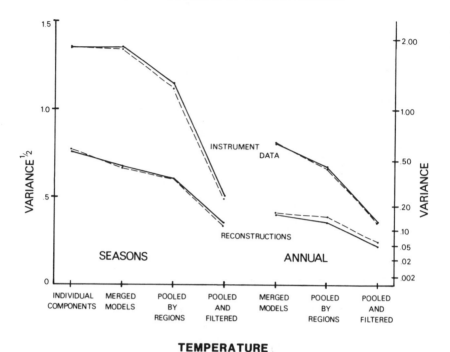

TEMPERATURE

Fig. 6.19. The standard deviations (*left*) and variances (*right*) of the instrumental data and reconstructions of temperature for the large grid (*solid lines*) and the small grid (*dashed lines*). The data were stratified, averaged, and plotted as in Figure 6.9.

If we assume a normal distribution, 95 percent of the data will have a range of four standard deviations. This amounted to 5.4° C for the seasonal grid-point temperature measurements and 3.0° C for the seasonal temperature reconstructions. When the seasonal temperature measurements were averaged for regions and the averages were filtered, the ranges decreased to 4.5° and 2.0° C, respectively. The annual averages had a range of 3.2°, the regional averages had a range of 2.7°, and the filtered data had a range of 1.5° C. The differences between grids for temperature were negligible. However, the ranges for the seasonal reconstructions showed a comparable but slower decline with increasing levels of pooling. The ranges for successive levels of pooling were 2.7°, 2.4°, and 1.4° C for the seasonal estimates and 1.7°, 1.5°, and 0.9° C for the annual values.

The standard deviations of precipitation (Fig. 6.20) declined more rapidly with the increased level of pooling than the temperature measurements. This difference simply may reflect the smaller scale of the precipitation variance compared to the temperature variance. As was true for temperature, the standard deviations of the precipitation

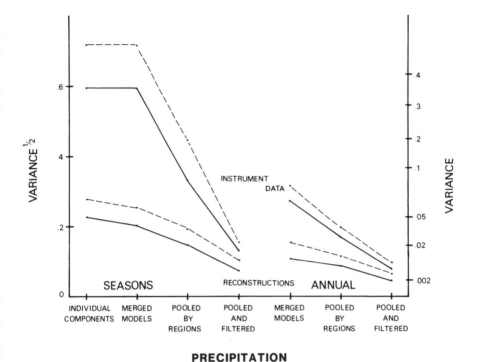

PRECIPITATION

Fig. 6.20. The same as Figure 6.19, for precipitation.

reconstructions were smaller and declined less rapidly than those for the instrumental data. The standard deviations of sea-level pressure also showed diminishing variance and increasing percentage of calibrated variance with increasing levels of pooling, but the data were pooled by eigenvector rather than by region. Since these data cannot be compared easily to the temperature and precipitation data, the plots were not included among the figures.

These differences in variance must be appreciated when evaluating the importance of different types of averages. A difference of a 1° change in an annual filtered estimate would be a truly remarkable event, as the 95 percent range is only 0.9°. A 1° change in the unfiltered annual averages would be important but not remarkable, as the 95 percent range is 1.7° C. A similar change in the seasonal merged models would be less important, as it would be compared to a 95 percent range of 2.7° C. In addition, the amount of variability in the reconstructions should not be evaluated as if they were actual measurements, because, by definition, the variances are a function of the amount of calibration, and the regression estimates for the dependent period always must be less than the variances of the dependent data.

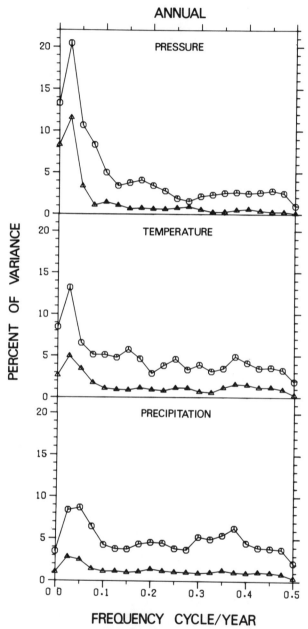

Fig. 6.21. Summed variance spectra for the first 5 PCs of the instrumental data (*circles*) and reconstructions (*diamonds*) expressed as a percentage of the variance in the instrumental data PCs. Temperature and precipitation data are for 1901–1961 from the large grids. The sea-level pressure data are for 1899–1961.

2. Variance Spectra

Spectral analysis (Blackman and Tukey 1958; Jenkins and Watts 1968) was used to decompose the variance in the first 5 PCs of each grid and variable into variance spectra. The lag-covariance method with the Hamming spectral window was used, and the spectra for the 5 PCs were summed to obtain a single spectrum for all of the grid-point data. Twenty-lag covariance and variance spectra were run on the annual twentieth-century instrumental record and reconstructions. Twenty-lag variance spectra also were obtained for the seventeenth through nineteenth century reconstructions. High-resolution, 90-lag spectra were run on the annual reconstructions for the entire length of record.

The 20-lag spectra for the twentieth century (Fig. 6.21) showed the reduced variance of the annual reconstructions compared to the instrumental data. The low-frequency peaks in the spectra of the instrumental data were reproduced in the spectra of the reconstructions, particularly for sea-level pressure and to a lesser extent for temperature. Some variance was reconstructed at higher frequencies, but no details of this structure are evident from the figure.

The differences in the variance estimates between the instrumental record and the reconstructions for each PC were tested for significance ($p \geq 0.95$), and these results are presented in Figure 6.22 along with coherency-squared values exceeding 0.5 and phase angles that were $2/8$ or larger (these values were considered important only when the coherence was 0.5 or larger). The frequencies that are shaded in the figure and are without phase-angle notation were most faithfully reconstructed. The largest PCs were more faithfully reproduced than the smaller PCs. The first 5 PCs of both sea-level pressure and temperature were well modeled. Only the first 3 PCs of precipitation were as well reproduced; the low frequencies of PC 4 and all but the lowest frequency of PC 5 were not reproduced in the precipitation reconstructions. There were only eight phase shifts associated with coherencies that were 0.5 or larger, so phase shift was not a problem. These results also confirmed the greater reliability of the low-frequency over the high-frequency reconstructions for sea-level pressure and temperature.

Three different spectral plots are superimposed in Figure 6.23 to reveal any differences between the reconstructions in the seventeenth–nineteenth centuries and the twentieth century. In addition, the 20-lag spectral estimates were separated and pooled into low-, middle-, and high-frequency classes (Table 6.9). Ratios were calculated to evaluate any differences between the two time periods. More variance was reconstructed in the seventeenth through nineteenth centuries than in the twentieth century for all three variables, and the differences were

ANNUAL
PRESSURE

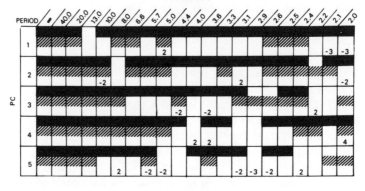

TEMPERATURE

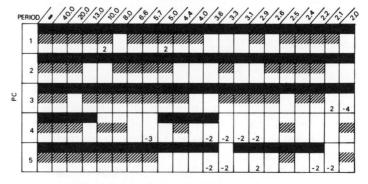

PRECIPITATION

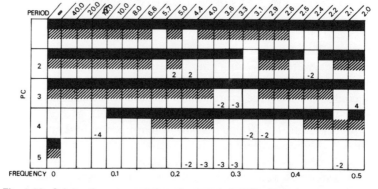

Fig. 6.22. Schematic representation of spectral similarity between reconstructions and instrumental data for the first 5 PCs of sea-level pressure, temperature, and precipitation (*large grids*) over the calibration period. *Solid bars*: frequency bands with no significant difference ($p < 0.95$); *shading*: frequency bands with coherencies of 0.5 or larger; *numbers*: possible phase shifts of 2/8 or greater. Phase shifts of this size are significant only when the coherency is 0.5 or larger and the phase angle is negative when the reconstructions lag behind the instrumental data.

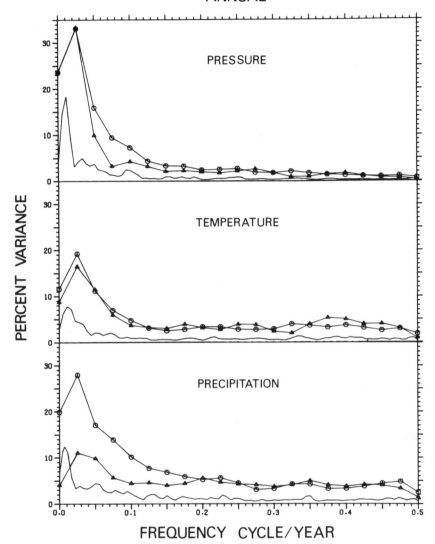

Fig. 6.23. Variance spectra summed over the first 5 PCs of the reconstructed data for sea-level pressure, temperature, and precipitation (*large grids*) expressed as a percentage of the total variance in the 5 PCs. *Triangles:* 20-lag spectra for twentieth-century data; *circles:* 20-lag spectra for seventeenth–nineteenth century data; *unbroken lines:* 90-lag spectra for the entire length of record.

Table 6.9. Percentage of Variance Summed over Three Broad-Frequency Bands and the Ratios for the 20-lag Spectra of the Annual Reconstructions

Variable, Time	Band Frequency (in cpy)			
	0.00 to 0.125	0.125 < to 0.300	0.300 < to 0.500	
	Period (in years)			
	Infinity to 8	8 > to 3.3	3.3 > to 2	Total
Pressure				
1899–1961	78	14	8	100
1602–1898	95	18	10	123
Ratio	0.82	0.78	0.80	0.81
Temperature				
1901–1961	49	23	28	100
1602–1900	57	20	25	102
Ratio	0.86	1.15	1.12	0.98
Precipitation				
1900–1961	39	31	30	100
1602–1900	97	34	31	162
Ratio	0.40	0.91	0.97	0.62

Note: Variance is expressed as a percentage of the twentieth-century reconstructions.

concentrated in the low frequencies. The variance estimates for the seventeenth through nineteenth centuries were higher at all three frequency bands for both sea-level pressure and precipitation, and they were lower at intermediate and high-frequency bands for temperature. This comparison of the variance spectra for the two time periods does not include an assessment of the variance associated with the twentieth-century mean changes. If the variance attributed to changes in the twentieth-century means had been added (as in Table 4.1), the total variance estimated for the twentieth century would have been larger for temperature and probably larger for sea-level pressure as well. Thus, the large anomalies reconstructed in these three variables reflect the fact that the regression analysis minimized variance in the twentieth century, so that if the calibration period was anomalous, large deviations would be reconstructed for other centuries.

It is notable that less variance was calibrated for precipitation than for the other two variables during the twentieth century and that the precipitation reconstructed had only 40 percent as much low-frequency variance and 62 percent as much total variance (Table 6.9) in the twentieth century as in the seventeenth through nineteenth centuries. This large change in variance probably reflects the deterioration of the reconstructions (Chap. 5, Sec. C) that would be expected for the precipitation because this variable was so poorly calibrated. These spectra confirm that the greatest proportion of variance was reconstructed at 0.1 cycles per year or smaller, representing time scales of 10 years or longer. The high-resolution (90-lag) spectra in Figure 6.22 indicate that the greatest variance was concentrated at frequencies less than 0.05 cpy, roughly corresponding to time scales of 30 to 200 years.

The greatest number of significant spectral peaks were noted in the low frequencies, and the smallest number were noted in the intermediate frequencies. This observation was consistent with the low number of spectral peaks observed at intermediate frequencies for the M chronologies that had been corrected only for the first-order autocorrelation (Fig. 4.3). The use of chronologies that are fully Autoregressive Moving Average (ARMA) modeled would be expected to resolve this problem.

7

Reconstructed Climatic Variations

There are four basic causes of climatic variations over time scales ranging from years to centuries (Lamb 1977). The first cause includes recurrent climatic changes, many of which appear cyclic, associated with solar output, the energy channeled toward particular parts of the earth, or tidal forces of the sun and moon. The second involves volcanic forcing of climate following great volcanic explosions that inject large amounts of dust into the stratosphere. It comprises all features of the global atmospheric circulation that appear to evolve over years and decades as the energy balance of the earth adjusts to large dust veils from the explosions. The third corresponds to changes generated by variations in the circulation and internal heat economy of the oceans. This category includes ENSO events, effects of pack ice, and variations in sea-surface temperatures. The fourth cause includes atmospheric features influencing transparency due to cloudiness, smoke and aerosols, and changes in atmospheric gases.

Considerable disagreement exists as to the relative importance of these factors to climatic variation. For example, Pittock (1978) takes a critical look at long-term sun-weather relationships and finds little convincing evidence of a real relationship. Eddy et al. (1982) find a real effect but believe it is too small to be significant in climate predictions. Stockton et al. (1983) find a weak but significant signal in drought reconstructions that relate to solar and lunar variations. Stuiver (1980) cannot find a significant relationship between climatic time series and a ^{14}C derived record of solar changes. Many workers believe that if there is a solar effect, it does not control a large part of the climatic variation.

Rampino and Self (1982), Kelly and Sear (1984), Bradley (1988), and Lough and Fritts (1987) examine the climate response to dust-veil forcing and report significant linkages to temperature, but all occur at lags of no more than 3 years behind the injection of dust into the atmosphere. Robock (1979) models the energy balance of the Northern Hemisphere and finds that forcing climate with volcanic dust over the past 400 years produces the best simulation. Forcing climate with sunspot numbers gives poor results. Porter (1981) finds that glacier variations in the Northern Hemisphere match the acidity record in polar ice cores and the frequency of volcanic eruptions in successive latitude belts. These correspondences suggest that volcanic aerosols produced during large eruptions could be a major factor in modulating climate on the decadal scale (also see Budyko 1969, 1974). Empirical evidence for a lag greater than 3 years has not been demonstrated to be statistically significant.

Namias (1982a, 1982b) relates sea-surface temperature teleconnections in the North Pacific to atmospheric conditions over North America. Andrade and Sellers (1989) and Schonher and Nicholson (1989) demonstrate that ENSO events can produce large precipitation anomalies in western North America. Lough and Fritts (1990) find a climatic response to the Southern Oscillation in these North American climatic reconstructions, but the response is short-lived and weak because it is missing the linkages from the Southern Hemisphere and from tropical regions. Catchpole and Faurer (1983) reconstruct the severity of sea ice in Hudson Strait and relate it to atmospheric circulation conditions. Melting of sea ice is associated with a westerly displaced Icelandic Low when mild southerly air is advected northward and an early dispersal of ice occurs.

A. Climatic Variations through Time

In this section the reconstructions are averaged over one or more regions and the average values filtered with the 13-weight low-pass digital filter described by Fritts (1976). The filter preserved the most reliable fluctuations throughout the 360-year time period which were longer than 7 years in duration and eliminated the least reliable reconstructions at time scales shorter than 7 years (see Chap. 6, Sec. F.2). Figure 7.1 shows the filtered averages for the temperature and precipitation grids as well as the averages for the western and eastern halves of the area (Table 3.1) (unless stated otherwise, all maps and plots are expressed as departures from the 1901–1970 means of instrumental data). For the twentieth century, the averaged reconstructions are plotted along with the instrumental record for the same areas. The high similarity between the instrumental data and the reconstructions, especially in the West, demonstrates the high calibration obtained for low-frequency information in the western regions. As noted earlier,

ANNUAL TEMPERATURE

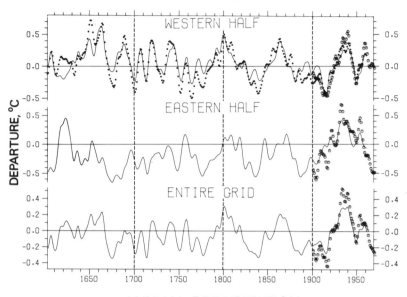

ANNUAL PRECIPITATION

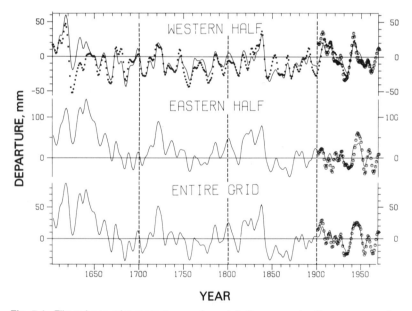

YEAR

Fig. 7.1. Filtered annual temperature and precipitation reconstructions expressed as anomalies from the twentieth-century values and averaged over the western half (regions 1–6), eastern half (regions 7–11), and the entire set of 77 and 96 data points. *Small dots*: filtered reconstructions from the small grids; *large dots*: filtered twentieth-century instrumental data.

Table 7.1. Decades with the Five Highest-ranking and Five Lowest-ranking Average Values of Annual Temperature

Grid	Highest Ranking					Lowest Ranking				
	1	2	3	4	5	1	2	3	4	5
77 Grid										
Total grid										
First year	1930	1800	1861	1656	1645	1833	1671	1880	1773	1908
Value	0.55	0.45	0.37	0.35	0.25	−0.55	−0.54	−0.48	−0.44	−0.42
Western half										
First year	1657	1797	1860	1681	1645	1908	1837	1948	1760	1611
Value	1.01	0.68	0.67	0.61	0.61	−0.64	−0.38	−0.34	−0.34	−0.34
Eastern half										
First year	1616	1931	1802	1948	1865	1671	1885	1833	1681	1772
Value	0.76	0.72	0.23	0.20	0.18	−1.08	−0.96	−0.96	−0.92	−0.89
46 Grid										
Total grid										
First year	1657	1681	1647	1717	1800	1695	1749	1760	1725	1909
Value	0.60	0.44	0.41	0.40	0.34	−0.34	−0.33	−0.31	−0.29	−0.28

Note: Departures in °C.

Table 7.2. Decades with the Five Highest-ranking and Five Lowest-ranking Average Values of Annual Precipitation

Grid	Highest Ranking					Lowest Ranking				
	1	2	3	4	5	1	2	3	4	5
96 Grid										
Total grid										
First year	1611	1632	1642	1718	1832	1842	1863	1751	1928	1682
Value	7.48	6.96	5.12	4.58	4.35	−3.20	−3.02	−2.45	−2.07	−2.02
Western half										
First year	1611	1831	1631	1718	1940	1842	1682	1856	1751	1663
Value	4.72	2.31	2.25	1.75	1.50	−3.32	−3.05	−3.01	−2.90	−2.54
Eastern half										
First year	1634	1611	1644	1718	1832	1863	1842	1930	1755	1768
Value	12.78	10.76	9.54	7.91	6.87	−3.84	−3.05	−2.42	−2.06	−1.98
52 Grid										
Total grid										
First year	1831	1907	1611	1717	1937	1621	1748	1663	1773	1856
Value	20.56	15.21	12.97	11.50	9.39	−31.28	−24.61	−23.68	−22.78	−19.08

Note: Departures in cm.

Table 7.3. Dates, Differences, and Rank of Most Significant 30-year Mean and
Standard Deviation Changes

| Beginning Year of Decade | Annual Temperature (in °C) | | | | | | Annual Precipitation (in cm) | | | | | |
| | Mean | | | Standard Deviation | | | Mean | | | Standard Deviation | | |
	Year	Difference	Rank	Year	Difference	Rank	Year	Difference	Rank	Year	Difference	Rank
1602												
1610												
1620												
1630	1637	0.24	9									
1640							1647	−3.92	3			
1650				1650	0.20	2						
1660	1667	−0.45	5				1663	−5.18	1			
1670												
1680				1682	−0.15	3				1688	−0.99	3
1690							1698	2.29	7			
1700												
1710	1717	0.27	7				1711	2.95	4			
1720				1724	−0.13	5	1728	−1.80	12	1729	−0.97	4
1730												
1740							1748	−2.76	5			
1750												
1760	1761	−0.26	8	1761	−0.11	7						
1770												
1780				1784	0.12	6	1783	2.05	9			
1790	1791	0.53	3				1797	1.82	11			
1800												
1810							1815	2.20	8			
1820	1821	−0.57	2							1821	1.45	2
1830												
1840							1842	−5.02	2			
1850	1850	0.42	6							1858	−1.90	1
1860												
1870	1877	−0.48	4				1873	2.38	6			
1880				1885	−0.15	4						
1890							1890	2.02	10			
1900										1904	0.80	5
1910	1918	0.63	1	1911	0.27	1						
1920												
1930												

Note: Reconstructed for the 77- and 96-grids.

many reconstructions for the eastern half of the continent were not reliable, especially those for precipitation, so the plots for eastern North American climatic variation must be interpreted judiciously and verified with other information from that area.

In addition, running means of the averaged reconstructions and their standard deviations were calculated for overlapping 10-year and 30-year intervals. Features of these data are presented in Tables 7.1–7.3 to help describe the large-scale, low-frequency changes in climate reconstructed over the United States during this time period (Fritts and Lough 1985). Tables 7.1 and 7.2 identify the five highest and five lowest nonoverlapping 10-year means. In addition, each 30-year mean was subtracted from the next nonoverlapping mean, and the differences between the means were tested for significance (Eq. A.18) to identify the times with the most rapid changes. The differences in variance were also tested using the F statistic (Table 7.3).

1. The Twentieth-Century Mean Changes

To relate the reconstructed history to modern data, it is helpful first to consider the features apparent at the beginning of the twentieth century. The greatest 30-year temperature change that was reconstructed was the warming beginning late in the nineteenth century. The greatest difference between 30-year means for any period was noted in 1918. The greatest variance change was an increase for the 30-year period beginning in 1911. The highest 10-year mean temperature for the entire 77-grid was reconstructed in the 1930s. It was 0.55° C higher than the 1901–1970 average value. Clearly, the largest-scale temperature change in the reconstructions occurred in the twentieth century, which is consistent with current notions about worldwide temperature changes (Schlesinger and Mitchell 1987; Hansen et al. 1988; Schneider 1989).

Nevertheless, none of the five highest average values reconstructed for the western half of the grid was from the twentieth century (Table 7.1). The twentieth-century temperature rise noted for the Northern Hemisphere (Jones et al. 1982) appears to have been delayed until the 1920s in western regions of the country (Boden et al. 1990). This cool period in the West early in the twentieth century coincides with the lowest temperatures noted by Paltridge and Woodruff (1981) for global sea-surface and air temperatures in coastal areas. Precipitation at the beginning of the twentieth century is, at first, high in the West and moderate in the East. In contrast, the 1930s were much drier, with some dryness evident in the 1950s and 1960s.

2. Seventeenth-Century Changes

Before the twentieth century the temperature departures reconstructed in western North America fluctuated around a mean that was slightly higher than that of the twentieth century, perhaps because of the low temperatures early in the century. Temperatures in eastern North America fluctuated around a mean below that of the twentieth century. The plots of temperature for the entire grid (Figure 7.1) and the results from Tables 7.2 and 7.3 show gradually rising temperatures during the early part of the seventeenth century, reaching high-ranking values in the latter half of the 1640s and 1650s. Contrasting temperature anomalies were reconstructed beginning in 1616 with warmth in the eastern half of the grid and moderate to cool conditions in the western half. The warm temperature reconstructions in eastern regions are not reliable, however, and need to be confirmed by independent information from the eastern regions. The second highest standard deviation change representing increasing variability was noted in 1650, and by 1667, temperatures over the entire grid were in rapid decline. The third largest standard deviation change from the 362-year-long reconstructed record occurred in 1682. Warming was reconstructed in the West but not in the East during the 1680s. The 46-station mean for the decade beginning with 1681 was second highest for that grid. Lower temperatures were reconstructed in the West for the 1690s, while warming was reconstructed in the East.

The data of Lamb (1970) and Hammer et al. (1980) indicate moderate peaks in volcanic activity in the 1630–40s and 1660–70s. Both periods were associated with warming in the West and cooling in the East. The large and persistent decline in global temperatures modeled in the late seventeenth century by Schneider and Mass (1975) to result from volcanic dust and hypothesized variations in the solar constant is not confirmed by these reconstructions of western North American temperature. It should be noted that the conclusions from this model were retracted and a revised model was constructed by Robock (1979) and compared to the temperature reconstruction of Groveman and Landsberg (1979). The simulations of Robock using volcanic dust for forcing resemble the dendroclimatic reconstructions of temperature for the seventeenth century more than the reconstructions of Groveman and Landsberg (1979), which were used in Robock's paper. The differences that do exist between the model and the dendroclimatic reconstructions could be due to model error, reconstruction error (because fewer rings were sampled from this time period), or differences between the climate of western and central North America and hemispheric-wide conditions used in the model (Fritts and Lough 1985).

The 96-grid precipitation reconstructions are valid only for the western regions. The first two decades in the seventeenth century were reconstructed to have been relatively wet in the West (Figure 7.1). The mean for the western half of the 96-station grid was higher for the decade beginning in 1611 than it was for any other 10-year period. The mean for the 52-station grid during that period was third highest-ranking, but beginning in 1621 the mean was lowest-ranking. The following decade was wetter. The mean during the interval beginning in 1631 for the western half of the 96-station grid was the third highest-ranking. Precipitation in the West was reconstructed to have been more or less average for the next two decades. The large differences between the two grids in the West from 1620 to 1640 suggest that errors may have been introduced by including a large number of distant stations from eastern regions. These differences did not appear to be large later in the record, so they may have been accentuated by the large errors in the chronologies attributable to the small number of trees available for these early time periods (see Chap. 2). Precipitation during the last half of the seventeenth century was reconstructed to have been near or below the twentieth-century average. Decade means for the intervals beginning in 1663 and 1682 were listed among the five lowest-ranking.

3. Eighteenth-Century Changes

Early in the eighteenth century, temperatures fluctuated around the twentieth-century mean figures with the decade beginning in 1717, reconstructed to have been the fourth warmest for the 46-station grid. The climate began to cool, and in about 1724 and again in about 1761 the standard deviation of the mean decreased by 0.13° and 0.11° C respectively (Table 7.3). The decade beginning with 1760 was especially cool in the West, but the cool decade beginning in 1772 was the only marked temperature anomaly of the century in the East. The reconstructed temperatures in the 1780s and 1790s began to warm throughout the country, reaching maximum values at the beginning of the nineteenth century. The variance of temperature also increased.

The mild temperatures in the West and a lack of large climatic anomalies were associated with low volcanic activity during the century (Lamb 1970; Hammer et al. 1980), with lowest values attained at midcentury. The temperature reconstructions for the West agree with the Robock (1979) model using volcanic dust forcing, except for the high temperatures at the end of the century. Lamb reports increasing dust veils toward the end of the century with the eruption of Laki in 1783, and Hammer et al. (1980) report that acidity peaks in Greenland ice increased near the end of the century. Lamb infers that dust veils might have been greater than estimated because of lower temperatures noted in Europe and New England for the 1760s, but the acidity profiles do not show the inferred dust fall during that period. The summer sea ice record of Catchpole and Faurer (1983) begins in 1751 and shows some years of very frequent ice encounters in the 1750s and mid-1780s, but there is a general decline in ice until 1810.

The agreement of high-frequency variations in these data was tested, using the DIFMAP program with the reconstructions (see Appendix 3) to average the reconstructions of summer temperature and precipitation for 18 years with high ice encounters. The average values for 1751–1870 were subtracted, and the differences at each grid point were tested for significance. The lack of any significant differences indicated that there was no agreement with the year-to-year variability in the Catchpole and Faurer data. The temperature reconstructions for eastern central Canada made from Guiot's (1987) high-frequency proxy records of temperature also were examined, although it was difficult to make comparisons using the plots presented in his paper. No marked agreement was evident with any data from eastern Canada. This result is not unexpected, as the Hudson Strait is at the margin of the sea-level pressure grid, far north of the temperature grid, and climate in that area is closely linked with conditions around Baffin Island and the North Atlantic (Catchpole and Faurer 1983; Guiot 1987).

The results from both precipitation grids are consistent over the century; dry conditions were reconstructed in the West at the beginning of the eighteenth century, although they were not as dry as in the previous century. Precipitation was reconstructed to increase, reaching the fourth highest-ranking decade averages beginning in 1717/1718 (Table 7.2). Although the fourth highest precipitation reconstruction, beginning in 1718 in the East, cannot be trusted, it did coincide with a time of high moisture reconstructed from tree rings in the southeastern part of the country by Cook et al. (in press). Drier conditions were reconstructed for the remainder of the century. The most extreme droughts were reconstructed in the decades beginning with 1748/1751 and 1773. Significant declines in the 30-year mean values and the standard deviations were noted during this time period. Possible dry decades were reconstructed in the East beginning in 1755 and 1768, and droughts were reconstructed in the East during the latter period (1769–1776) by Cook et al. (in press).

It is notable that the period from 1650 to 1720 (Eddy 1976, 1977)—the Maunder minimum, when sunspot numbers were especially low—does not stand out as a period of greatly different climatic conditions (see the simulations of Robock [1979] for temperature estimates using smoothed sunspot numbers for climatic forcing). Two features are evident in

Figure 7.1 during this time period. The first is a cooling trend in the West with a warming trend in the East interrupted by cooling in the 1670s, 1695–1704, and 1710s. The second is the presence of four to six droughts interrupted by periods of more average precipitation, depending on how one interprets the figure. The oscillations in temperature and the droughts are consistent features in both the large- and small-grid reconstructions, and no intervals with high precipitation are evident. These same features can be seen during other time periods, but with a lower-frequency pattern apparently superimposed on the oscillations.

4. Nineteenth-Century Changes

The second highest 10-year mean for the 77-station temperature grid was reconstructed in 1800 (for the 46-station grid the mean value was fifth ranking). The temperature trend was downward for 40 years following the maximum at the beginning of the nineteenth century. The temperature decline was most rapid around 1821. The coldest decade reconstructed for the entire grid began in 1833, and it was colder in the East than in the West. Temperatures began to rise in the 1840s, and in 1850 the temperature change became most evident. The next highest-ranking 10-year mean value was reconstructed in 1860 for the western half of the 77-station grid and in 1861 for the entire grid. The decline in temperature that followed was most rapid in 1877. The average temperature for the large grid reached its third lowest 10-year mean value in 1880, but the greatest cooling was reconstructed in the East, the area of the most unreliable reconstructions. There was a significant decline in the reconstructed variance around 1885. Temperatures remained low or moderate until the turn of the century.

The data of both Lamb (1970) and Hammer et al. (1980) show high volcanic activity accompanied by large dust veils in the first half of the century, when low temperatures were reconstructed for large parts of the United States. In fact, the grid-wide average reconstructed temperature follows the volcanic model of Robock (1979) more closely than the reconstructions of Groveman and Landsberg (1979), which used a heterogeneous set of data for this time period (Fritts and Lough 1985). The data of Catchpole and Faurer (1983) and Guiot (1987) show that declining temperatures were most marked in the early part of the century. As Lough (in press) demonstrates, however, the extremely cold temperature anomalies associated with the eruption of Tambora are not associated with large temperature anomalies in western North America and the North Pacific, although the temperatures were declining during that time period. The warming of the 1840s–1860s, followed by cooling, also was associated with a period of low volcanic activity followed by a period of high volcanic activity. Again, the average temperature reconstructions follow the volcanic model of Robock closely.

As temperature declined, precipitation increased during the first four decades of the nineteenth century, although dry conditions were reconstructed at the beginning of the second decade. The highest and second highest 10-year averages of precipitation were reconstructed for the 52-grid and western 96-grid starting with 1831. This decade clearly appears to have been an extremely wet one for a large number of stations. Much drier conditions were reconstructed for the remainder of the century. The decline in moisture beginning in 1842 was the second-ranking change, and the decline in standard deviation starting in 1858 marked the greatest variance change that was reconstructed for the 96-grid data. The 10 years beginning with 1842 and 1863 were the first and second lowest-ranking values from the 96-grid, and those beginning in 1856 were the fifth lowest-ranking from the 52-grid reconstructions. The drought was somewhat ameliorated by 1873 and again by 1890, as is evident in the 30-year mean changes. The variance of precipitation remained low until the twentieth century.

East-west seesaw patterns of temperature variation, like those shown in the temperature eigenvector patterns (see Chap. 4 and Diaz and Fulbright 1981), can be noted. In addition, Lough and Fritts (1987) report opposite temperature responses between the central United States and the Pacific Coast associated with dustveils from low-latitude volcanic eruptions.

Cook et al. (in press) report that 1814–1822 was characterized by severe drought in the East. The annual and seasonal reconstructions were examined with DIFMAP for this particular time period. Though the average annual reconstructed precipitation in the West was near the twentieth-century value or below for that time period (Fig. 7.1), warm dry conditions were reconstructed in the Far West and central Plains in spring and summer (Fritts and Shao in press). Cook and his colleagues also report that 1827–1837 was one of the wettest periods in eastern United States, a finding that is consistent with the aforementioned reconstructions averaged over the eastern and western regions.

B. Regional Variations and Grid Differences

The variations in reconstructed temperatures in the six western regions often were similar for the small and large grids (Fig. 7.2), although there were specific time periods and localities in which different temperature patterns were evident. The most marked inconsistencies

TEMPERATURE

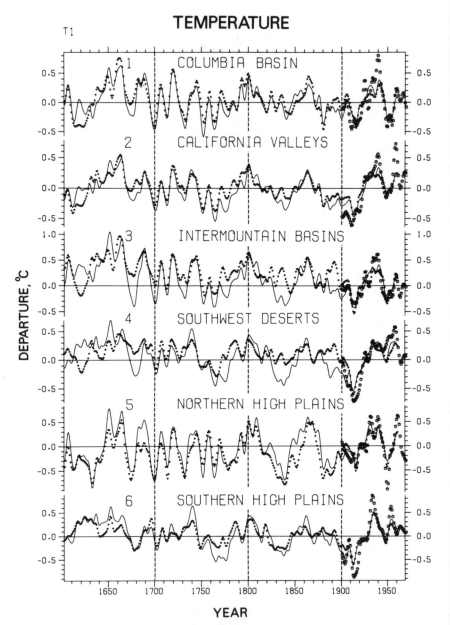

Fig. 7.2. The filtered 46-grid (*line*) and 77-grid (*small dots*) annual temperature reconstructions, averaged over all grid points in each of the six western regions and expressed as anomalies from the twentieth-century values. The *larger dots* in the twentieth century represent the filtered instrumental data.

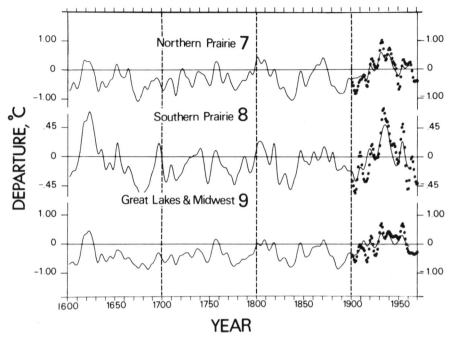

Fig. 7.3. Same as Figure 7.2, for regions 7–9.

between the two grids were noted in regions 1–4 in the 1620s, 1670s, 1770s, 1830s, 1840s, and 1890s. The large low-frequency temperature variations that were evident in all six regions during the seventeenth century diminished in magnitude and period length during the eighteenth century. Low-frequency temperature fluctuations became more pronounced in all regions after 1800 and culminated in the West early in the twentieth century with cool conditions that were then followed by regional warming.

The only reconstructions for regions 7–9 from the central and eastern United States (Fig. 7.3) came from the large grid. During the 1610–1620s, temperatures were reconstructed to rise above the 1901–1970 averages, but then they declined and remained mostly below this average until the beginning of the nineteenth century. The coldest periods were in the 1670–1680s, 1770–1780s, 1820–1830s, and 1880–1890s.

Low-frequency changes in precipitation for the western regions (Fig. 7.4) show more differences between the two sizes of grids than do those for temperature. The greatest differences between grids were reconstructed for region 5, the northern High Plains. Precipitation was

PRECIPITATION

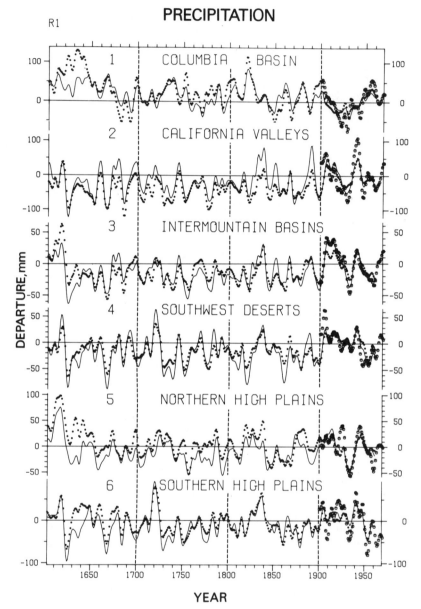

Fig. 7.4. Same as Figure 7.2, for the 57-grid (*line*) and the 96-grid (*small dots*) annual precipitation reconstructions.

reconstructed to have been higher for the 57-station grid than for the 96-station grid in regions 1, 5, and 6 early in the seventeenth century. Precipitation in the eighteenth century was often somewhat higher for the 57-grid reconstructions than for the 96-grid reconstructions in regions 1 and 5, the most northern regions. The 96-grid reconstructed precipitation was often lower for region 2, the California valleys, during all three centuries. The general patterns through time for the two grids were reconstructed to have been similar throughout the western regions, however, except that the precipitation variations in the north-ernmost regions, 1 and 5, were out of phase with the variations reconstructed in southern regions. These results are consistent with the first three eigenvector patterns (Fig. 4.5), which all show opposite variations between the Pacific Northwest and the Southwest.

Low-frequency variations in precipitation varied among some regions. Region 1 was reconstructed to have been generally wet with drier conditions near the end of the century. The droughts of the 1620s and 1660s were more evident in the other five regions, and it was generally drier than in the north. Similar features were apparent in the eigh-teenth and nineteenth centuries, but the very low-frequency variations were not as marked. The percentages of precipitation reconstructed for region 1 were more often above the twentieth-century mean values than were the percentages reconstructed for regions 2–6. Low-frequency changes were more evident in the twentieth century for all western regions.

Few precipitation estimates for regions 7–9 had acceptable statistics; they were, on the average, below those for precipitation in the western regions and for temperature in regions 7–9. The time-series plots for precipitation in regions 7–9 have not been included, but the ten best precipitation reconstructions were averaged from a region that now includes South Dakota, Nebraska, Kansas, Oklahoma, Iowa, and Mis-souri. The averages were filtered, and the results were plotted with the filtered temperature estimates for the same selected grid points (Fig. 7.5) (Fritts 1983).

The two plots in Figure 7.5 were compared to the reconstructions of droughts by Stockton and Meko (1975), although these two data sets emphasize different frequencies and regions (the Stockton and Meko reconstructions used higher-frequency information on drought-severity indices for the entire western region) and used a slightly different tree-ring chronology data set (see Chap. 2 for a description of the data sets). The years reconstructed to have the highest temperatures and lowest precipitation were compared to the years reconstructed to have a low drought-severity index.

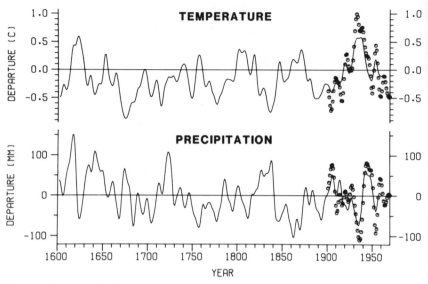

Fig. 7.5. Filtered annual temperature and precipitation reconstructions averaged from nine temperature stations and nine precipitation stations in South Dakota, Nebraska, Kansas, Oklahoma, Iowa, and Missouri.

The following inferences were drawn: (1) The filtered reconstructions of temperature from the ten stations in the central Plains states suggested that the climate was generally cooler than the 1901–1970 average values except for warm periods in the 1610–1620s, 1650, 1740s, 1760s, 1800–1810s, mid-1840s, and 1860–1870s. (2) Only the 1620–1630 interval was reconstructed as warm as it was in the 1930s. However, the 1610s and 1630–1640s were reconstructed to be wet, whereas the 1620s were dry. An apparent drought was interspersed between two wet periods, and its duration appeared to be shorter than the drought of the 1930s. (3) For the other intervals of above-average temperature, the precipitation was reconstructed to have been low only in the 1750s, 1810s, 1840s, and 1860s.

These results agree with those of Stockton and Meko (1975) in that when moisture and temperature were considered together, drought conditions between 1700 and 1900 appeared not to have been as severe as the drought of the 1930s. The four driest 3-year periods reconstructed before the twentieth century by Stockton and Meko (1975) for the entire West are listed in order of their importance as: 1845–1847, 1863–1865, 1755–1757, and 1734–1736. Although 3-year variations averaged over the entire West would not necessarily be reflected in the filtered reconstructions from the Plains states (Fig. 7.5), three out of four were within periods reconstructed to have low precipitation with temperatures near or above the twentieth-century average values. The Stockton and Meko chronologies did not extend into the seventeenth century, so they had no information on the drought of the 1620s. In light of the differences in the variables that were calibrated and the methods used by the two analyses, the agreement between these two data sets is noteworthy. The plots of Figure 7.5 suggest that droughts before the twentieth century appear to have been less severe than the drought of the 1930s, perhaps because of cooler temperatures, but some of the droughts reconstructed before the twentieth century appear to have been longer-lasting. The fact that both investigations came to this conclusion adds to its credibility.

The drought reconstructed by Cook et al. (in press) in 1769–1776 for the East appears to be present in the central Plains and appears to have been accompanied by cool annual temperatures in that area. The warm dry interval followed by a cool moist interval reported by Cook and his colleagues is also apparent in Figure 7.5, although the maxima and minima of the filtered series do not coincide exactly. The lack of commonality in the occurrence of multiyear periods of drought and wetness reported by Cook et al. (in press) is also a feature of the reconstructions from western North America. The filtering is an attempt to extract the common signal that is present over wide areas. Specific reconstructions in specific areas are better examined by mapping the annual or seasonal results for exact time periods (Fritts and Shao in press).

C. Spatial Variations in All Three Variables

The calibration and verification statistics suggest that the most reliable reconstructions of spatial patterns are those that are (1) combinations of several models, (2) the averages of all four seasons, and (3) the averages of reconstructions from a number of years. The following discussion examines the spatial patterns averaged over a number of years. Unless specified otherwise, each map is a 10-year average expressed as an anomaly from the 1901–1970 mean values. The maps for all independent reconstructions for the 1602–1900 period were examined first to help answer the question, How different are the climatic anomalies described in each map from the twentieth-century measurements? The 9 years associated with the most rapid temperature changes (Table 7.3) were used to divide the 1602–1960 interval into ten successive periods with significantly different mean temperature regimes. The mean patterns and the changes associated with each of these periods

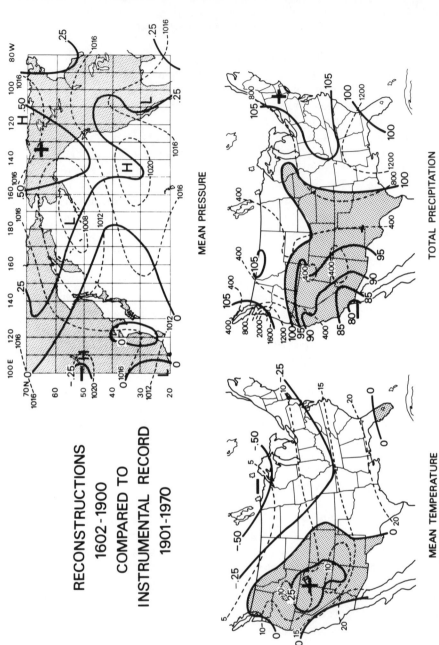

RECONSTRUCTIONS
1602-1900
COMPARED TO
INSTRUMENTAL RECORD
1901-1970

MEAN PRESSURE

MEAN TEMPERATURE

TOTAL PRECIPITATION

Fig. 7.6. The reconstructed mean sea-level pressure, mean temperature, and total precipitation anomalies for the 1602–1900 period expressed as departures from the 1901–1970 instrumental average. *Dashed lines:* mean values of the instrumental data for 1901–1970; *shaded areas:* warm and dry anomalies.

were then considered along with a description of the decade-by-decade changes observed within each period. Comparisons were made between past and present climatic variations with reference to present-day state and national boundaries.

1. The 1602–1900 Mean Patterns

The interval from 1602 to 1900 is sometimes referred to as a period falling within the Little Ice Age (Lamb 1977; Grove 1988) or Neoglaciation (Porter and Denton 1967). The means from this time period can be compared to the means of the twentieth century to express the most recent portion of the Little Ice Age climate in terms of the twentieth-century climatic conditions.

Seventy-five percent of the reconstructed sea-level pressure grid points for 1602–1900 (Fig. 7.6) were above the twentieth-century values, and the largest anomaly was greater than 0.5 mb for large portions of eastern Alaska and northwestern Canada. Such an anomaly for a 299-year period is a substantial departure from the twentieth-century mean values, considering the number of years that were averaged (see Chap. 6, Sec. F.1). It can be inferred from the positive anomaly in sea-level pressure that greater subsidence and a strengthened flow off of the polar ice pack (Bryson and Hare 1974; Kelly et al. 1982) may have occurred in the Little Ice Age. The Aleutian Low appears to have been displaced west of its twentieth-century mean position. The Icelandic Low also may have been displaced westward with an anomalous southward flow bringing cooler temperatures in the East, although the center of the Icelandic Low was largely outside the area of this study. An anomalous low that might indicate a weakened Central Asian High also can be noted, but the low reliability for reconstructions of sea-level pressure in the far western regions indicates that additional independent evidence is needed before this observation can be taken seriously.

The mean temperatures were reconstructed to have been higher, not lower, than the twentieth-century average for the western mountain region of the United States and lower elsewhere, especially in the Great Lakes region. Precipitation was reconstructed to have been lower than the twentieth-century average from California to western Wisconsin and southward to Kansas, Oklahoma, and east Texas. Precipitation was reconstructed above average from the Pacific Northwest to the Great Lakes and for the entire eastern half of the country. However, the error terms for most of the eastern precipitation estimates were too large to make a conclusion about precipitation in this region without other independent evidence.

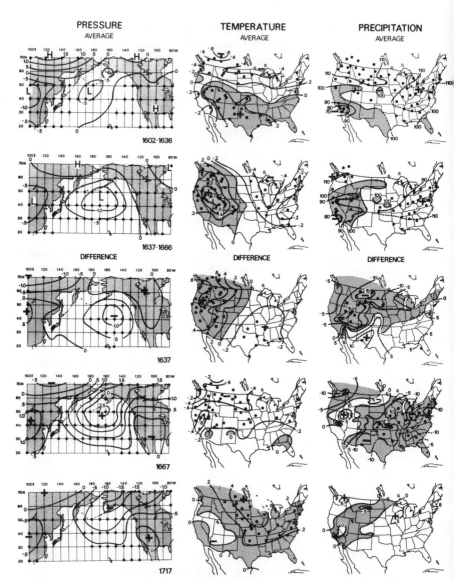

Fig. 7.7. Reconstructed climatic patterns separated by the most marked and significant changes in the 30-year mean temperatures (Table 7.3). *Rows 1 and 2*: averages in sea-level pressure, temperature, and precipitation for 1602–1636 and 1637–1666 shown as departures from the 1901–1970 mean values; *row 3*: differences between 1602–1636 and 1637–1666; *Rows 4 and 5*: successive difference maps. *Dots*: data points with significant anomalies ($p \geq 0.95$); *shaded areas*: warm and dry anomalies.

These particular maps are difficult to interpret because they are generalized as annual values over a very long time period. One interpretation is that the blocking of low-latitude storms off the North American coast was more frequent, deflecting them south of their twentieth-century mean position in the central North Pacific and then guiding them in a northeastward direction as they entered the Pacific Northwest. This pattern would produce an anomalous northward airflow in the Southwest with warmer temperatures than in the twentieth century. The anomalous high pressures reconstructed in the Arctic might be associated with a southerly displaced Arctic Front (Wendland and Bryson 1981), with a greater frequency of cold Arctic air outbreaks over the central Plains and Great Lakes. These data also suggest that fewer storms and associated fronts may have moved directly inland across southern California, the southern Rocky Mountains, and the southern Plains. This pattern would be consistent with the lower precipitation that was reconstructed there. Precipitation along the western United States-Canadian border, however, and perhaps farther east, was reconstructed to have been higher than the twentieth-century value, which would be consistent with more storms, greater frontal activity, higher precipitation, greater snow accumulation, and glacial advances in that region. If there was a Little Ice Age record in these data, glacial advances during this time period were more likely associated with an increase in annual precipitation than with a decrease in annual temperature. No decline in the annual average temperature over the West can be noted, but seasonal patterns in temperature were evident, with cooler winters and springs reconstructed for many western areas. Though the seasonal variation is beyond the scope of this volume, the seasonal reconstructions can be computed easily using the DIFMAP program described in Appendix 3.

The following anomaly maps have been corrected for the twentieth-century mean pattern. Thus, the seventeenth–nineteenth century anomalies will appear to have the patterns shown in Figure 7.6 embedded in them because the twentieth-century patterns were so different. In some cases, differences between two reconstructed maps were calculated to avoid this problem and to bring out the relative changes.

2. 1602–1636

The mean patterns of sea-level pressure, temperature, and precipitation reconstructed for 1602–1636 are shown in the first three maps in row one of Figure 7.7. The first two rows of maps in Figures 7.7 and 7.8 show the mean patterns, and the third row is the difference between these two periods. The remaining maps in both figures are differences.

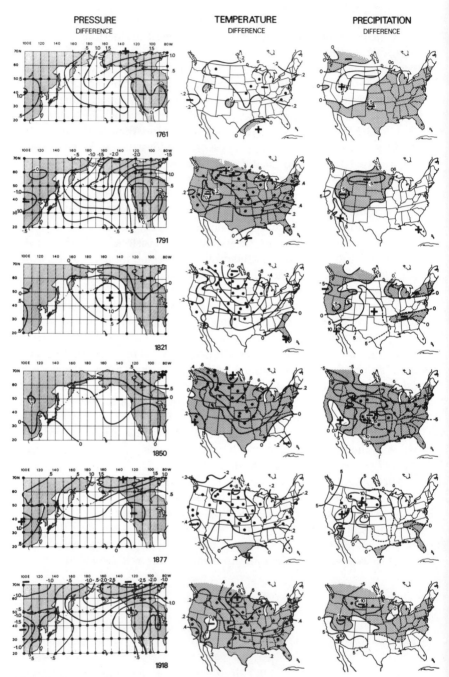

Fig. 7.8. Same as Figure 7.7, for successive difference maps 1771–1918.

The period 1602–1636 was reconstructed to have been anomalously cool in the northern states, with higher reconstructed sea-level pressures than in the twentieth century throughout the United States and the eastern North Pacific. Sea-level pressure was lower in the western North Pacific and higher over northern Siberia, but the latter reconstructions are in an area with questionable instrumental pressure data for the early twentieth century (Trenberth and Paolino 1980) and probably the northern Siberian reconstructions are unreliable.

Temperatures were reconstructed to have averaged as much as 0.7° C below the twentieth-century means for the northern half of the United States and southwestern Canada, including the Pacific Coast of southern California. Above-average temperatures were reconstructed for the Great Basin through the Southeast to the Atlantic Coast. Precipitation was reconstructed to have been 10 percent or more above the twentieth-century mean values in the Northwest, northern Plains to the central Great Lakes, Ohio Valley, and central Appalachia. The California coast, the central Great Basin, and the entire Southwest to central Texas were the only regions reconstructed to have been drier than in the twentieth century. For instance, precipitation was reconstructed to have been 68 percent of the twentieth-century values for Needles, California.

Figure 7.9 displays the decade maps of sea-level pressure, temperature, and precipitation for the first half of the seventeenth century. The first four decades were distinctive for their very large changes in climate. The maps for 1602–1610 are close to the embedded pattern shown in Figure 7.6. The Aleutian Low strengthened and extended eastward into Canada in 1611–1620 and intensified in 1621–1630 as lower pressures expanded into the North American Arctic. A blocking high is suggested by the pattern along the western coast of the United States; such a block could have enhanced the onshore flow in Canada and blocked storms from entering the United States along the coast of California. Temperatures appear to have been warm in the southern and east-central portions of the United States, and drought appears to have spread through the Southwest and central Rocky Mountains and slightly exceeded the embedded pattern (see Fig. 7.6). One synoptic interpretation of these patterns suggests that storms more or less followed the average pattern of the seventeenth through nineteenth centuries in 1602–1610.

In 1611–1620, sea-level pressures were lower over Alaska and Canada, the central and southern portions of the United States were warmer, and precipitation was above average for large parts of the country. These data may be interpreted as a low-pressure trough developing over the West in 1611–1620, increasing the number of storms,

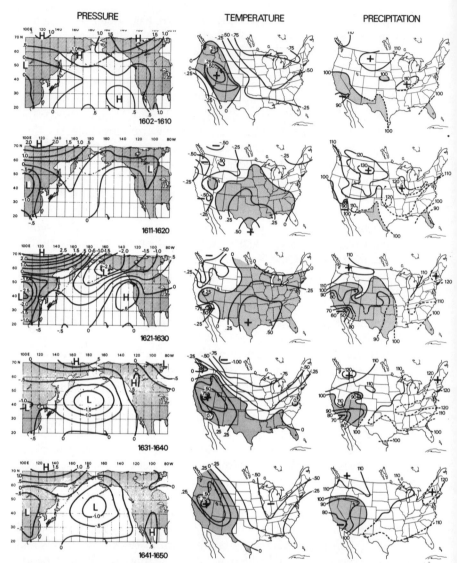

Fig. 7.9. The average reconstructed climatic patterns in sea-level pressure, temperature, and precipitation expressed as departures from the 1901–1970 mean values for 1602–1610 and subsequent decades in the first half of the seventeenth century. *Dashed lines*: areas of unreliable reconstruction; *shaded areas*: warm and dry anomalies.

producing high precipitation over much of the United States, and enhancing southerly flow, bringing warmth to the southern and central part of the country. The large anomalies reconstructed in 1621–1630 suggest a northerly displaced and strengthened storm track with lower subarctic pressures, reduced numbers and intensities of cold air outbreaks, and more advection of warm air from the tropics. The high pressure along the United States Pacific Coast may reflect a block in the West, with associated dry conditions and southward airflow that resulted in cooling along the coast.

The eruption of Vesuvius in December 1631 injected a dust veil with an index of 600 into the atmosphere, but it was not a low-latitude eruption and may not have affected the climate greatly in this area (Lough and Fritts 1987). By 1631–1640 the Aleutian Low had moved into a more southerly location, higher pressures had returned to western Canada, and storm activity may have been reduced in that region. Lower anomalies in pressure that were reconstructed over Hudson Bay could be interpreted as a westerly displaced or intensified Icelandic Low. Such a circulation pattern would strengthen the southward flow of cool air into the eastern United States, causing temperatures to decline in that area. The low moisture reconstructed in the Southwest was about the same as the embedded pattern.

3. 1637–1666

The eruptions of Komagatake, Awu, Santorini, Katla, and Usu produced dust veils in 1640, 1641, 1650, 1660, and 1663, respectively. Awu, in 1641, was the largest, with a dust-veil index (DVI) of 1000, and it was the only event at low latitudes. The anomalous lows that were established over the central North Pacific and over Hudson Bay during 1631–1640 dominated the circulation through at least 1666 (Fig. 7.7, row 2). Counterclockwise circulation about these lows brought warming to the western states and cooling to the East. Temperatures rose as much as 0.65° C in the Columbia Valley and southwest Canada and declined as much as 0.44° C for eastern areas, especially in the Mississippi drainage.

Perhaps storms moved into the Pacific Northwest a little more frequently than before, favoring precipitation along the Canadian and United States border and a return to the 1602–1900 embedded pattern (Fig. 7.6). The difference between the 1602–1636 and 1637–1666 periods (Fig. 7.7, row 3) shows falling pressures over the eastern North Pacific and rising pressures over western North America, warming in the West and cooling in the East with declining precipitation for large portions of the western and northeastern United States. Precipitation increased from southern Arizona to the Atlantic Coast (the difference maps eliminate the anomaly associated with the embedded 1602–1900 mean pattern).

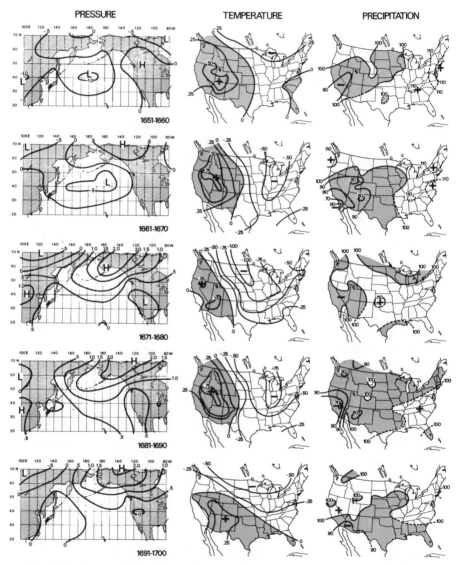

Fig. 7.10. Same as Figure 7.9, for the last five decades of the seventeenth century.

The decade maps for 1641–1670 (Figs. 7.9 and 7.10) show very similar sea-level pressure and temperature patterns. The intensities of the low-pressure anomalies gradually weakened, however, and the area of low precipitation expanded in a northeasterly direction, producing drought throughout much of the Great Plains region.

4. 1667–1716

Four great volcanic eruptions were noted between 1667 and 1716: Tarumai (1667), Gamkonora (1673), Serua (1693), and Fuji (1707). The largest was Gamkonora, with a DVI of 1000; both Gamkonora and Serua were low-latitude events that could have affected climatic variations in the area (Lough and Fritts 1987).

Sea-level pressures were reconstructed to have risen above values for the prior period over most of the North Pacific and the adjacent continents (Fig. 7.7, row 4). The rise was greatest (2.7 mb) south of the average annual position of the Aleutian Low (Fig. 7.6), implying a weakening of this low and a possible strengthening of the North Pacific High, as well as a strengthening of the high in the Klondike (Fig. 3.3). Sea-level pressures over the United States West were slightly lower. Temperature declined throughout most of the United States by as much as 0.61° C in California, and precipitation declined by 15 percent in the southern states. Precipitation was reconstructed to have increased only along a narrow band stretching from California northward across the intermontane basins, Montana, and North Dakota.

All of the decade maps for sea-level pressure from 1671 through 1720 show anomalously high pressure, first over Alaska and the west Canadian Arctic and then centering over the Gulf of Alaska (Figs. 7.10 and 7.11). This feature had appeared only once before, in the map for 1602–1610, but was to become more common in the future. A weak negative anomaly of sea-level pressure was also apparent over southwestern North America. Although the average temperatures in the West were cooler than before, the weak embedded positive temperature anomaly of the 1602–1900 period was present in all cases. The precipitation anomaly was more variable. The embedded pattern was less apparent, and moderately dry conditions alternated with more moist intervals during this time period (Figs. 7.1 and 7.2).

Fewer storms may have been associated with the weaker Aleutian Low, and higher sea-level pressures were reconstructed over northern North America during the 1670s through 1710s (Figs. 7.10 and 7.11). Storms may have been deflected south over the North Pacific so that more storms entered along the southern California coast and traveled northeast into the northern prairie states, producing the belt of

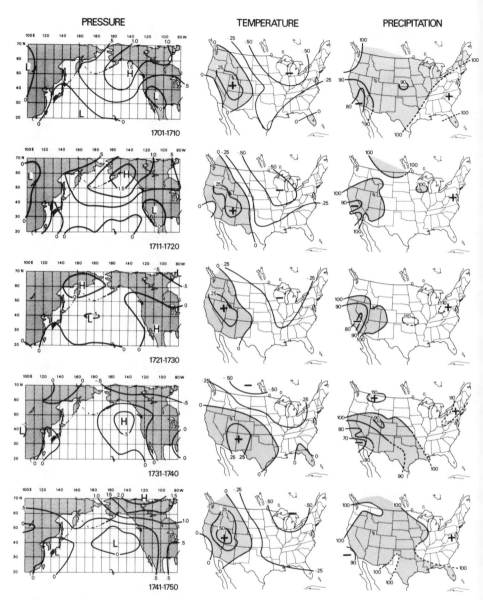

Fig. 7.11. Same as Figure 7.9, for the first five decades of the eighteenth century.

enhanced precipitation there (Fig. 7.7). Cooler air may have been advected from areas of high-pressure anomaly in Alaska and the Canadian Arctic into the United States, as indicated by the colder temperatures reconstructed there. Substantially fewer storms may have moved through the Great Plains and into the East.

The 1667–1716 interval began approximately twenty years into the Maunder minimum and ended at about its close. The reconstructions suggest that fewer or less intensive storms moved through the area, except in the band of enhanced precipitation centered in the Great Basin. The interval was cooler than it had been earlier, but the cooling pattern was not as extreme as in the Robock (1979) model, which uses sunspot numbers and volcanic activity as forcing functions of climatic variability.

It is possible that this period's persistent cooling could be associated with the volcanic activity (Lamb 1970; Robock 1979). Only a response with a lag of 1 to 2 years was detected by Lough and Fritts (1987). It involved only low-latitude eruptions with warming in the West and cooling in the central and eastern parts of the country. Some of the high-frequency temperature fluctuations shown in Figures 7.1 and 7.2 for this time period were related to dust-veil effects, but it is by no means certain that dust veils produced the general pattern of cooling.

Kocharov (1986: Fig. 11) reports tree-ring evidence from the USSR showing that the atmospheric radiocarbon was higher during the Maunder minimum, and values fluctuated from one cycle to the next during this time period. These data suggest that there was more modulation by the cosmic ray flux during the Maunder minimum, which could have influenced the climatic variations through this time period. Interesting spatial variations in reconstructed temperature and precipitation were noted; they appeared to be associated with the ^{14}C modulations reported by Kocharov (1986) for this interval, but judgment is withheld on the validity of a possible relationship until the data of Kocharov can be confirmed by independent workers in other countries. Some interesting and marked variations in climate were reconstructed during the Maunder minimum, but the individual effects and importance of the several possible causes of high-frequency climatic variation could not be identified with a conclusive analysis. Perhaps the matter can be resolved when more proxy data, including longer and better-replicated tree-ring chronologies can be assembled for this time period. Until that time, the reconstructed seasonal and annual variations from this study can be extracted from the available disk (App. 3) and analyzed with respect to new models of possible climatic forcing functions during this time period.

5. 1717–1760

Three notable volcanic events were identified during the 1717–1760 interval: Katla (1721), Tarumai (1729), and Katla (1755). The largest was the second eruption of Katla, with a DVI of 1200; none was a low-latitude eruption. The volcanic dust levels were generally low during this period (Lamb 1970). The climatic changes around 1717 (Fig. 7.7) are the reverse of those occurring early in the Maunder minimum. Sea-level pressure declined over large areas of the grid and increased slightly over southwestern North America and northern Siberia. The Aleutian Low appeared to strengthen, the North Pacific and Klondike highs (Fig. 3.3) weakened, and more storms may have moved inland across Alaska, northwestern Canada, and the American Arctic. The reconstructed temperatures during 1717–1760 were as much as 0.54° C higher than the former period, and the changes were statistically significant throughout the northern Plains to the East Coast of North America. Temperatures from Oregon to northern Arizona and New Mexico were unchanged or lower.

The band of rising precipitation in the West shown in the 1667 difference map became a band of declining precipitation in the 1717 map, but the percentage changes were small and the differences insignificant. Fewer storms may have entered along the California coast, and the circulation returned to 1637–1666 conditions.

In 1721–1730 (Fig. 7.11) the climatic anomalies were weak. By 1731–1740 a well-developed high was apparent over the eastern North Pacific, and low-pressure anomalies were reconstructed in the Canadian Arctic. The temperature and precipitation anomalies were close to the embedded patterns. A weak anomalous low in 1741–1750 replaced the high in the eastern North Pacific, and the Klondike High increased in strength. Drought expanded eastward but decreased in intensity in the Far West. By 1751–1760 the sea-level pressure anomalies had diminished along with moderate warming and dry conditions in the West. The overall climatic patterns of 1717–1760 were small and variable from one decade to the next. They resembled the means of the twentieth century more than the means of the 1602–1900 embedded pattern. No large-scale climatic patterns were reported in the eastern United States by Cook et al. (in press) during this interval.

6. 1761–1790

Important volcanic events in the 1761–1790 interval were Hekla (1766) and Cotopaxi (1768) and Eldeyjar, Jokull, and Asama (all in 1783). Cotopaxi was a low-latitude eruption, and Eldeyjar was the largest, with a DVI of 2300. The eruptions in the 1760s are associated with general cooling, and those of the 1780s are associated with warming (Fig. 7.1).

The difference in the average temperature between 1717–1760 and 1761–1790 (Table 7.3, Fig. 7.8) was the second smallest noted. The sea-level pressure patterns were the reverse of the previous period, and the maps of the differences appear to be weak versions of the 1667

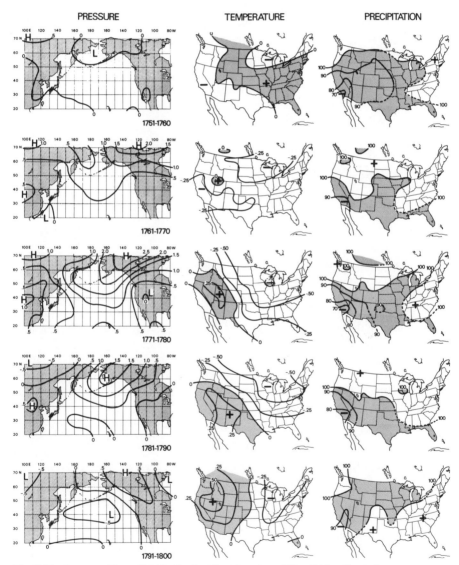

Fig. 7.12. Same as Figure 7.9, for the last five decades of the eighteenth century.

maps. A rise in sea-level pressure was reconstructed for all but the western and central United States. The greatest rise was in the area of the Klondike High. The Aleutian Low may have become weaker and southerly displaced. Less storm activity may be implied in the extreme Northwest along with southerly displaced storm tracks across the North Pacific. More cold air from the Arctic may have been advected into the United States.

The reconstructed differences in temperature were less than $-0.2°$ C, and few were significant. Precipitation increased somewhat from the Great Basin to the Dakotas, while precipitation decreased elsewhere in the grid. No precipitation difference was significant. Increasing numbers of storms may have entered the United States along the Oregon and California coast and moved inland, bringing increased moisture to the West as far as the northern Plains states. Similarly, fewer or weaker storms are implied elsewhere on the map.

The sea-level pressure maps for all three decades, 1761–1790 (Fig. 7.12), show a return to high-pressure anomalies in the North American Arctic, an Aleutian Low of diminished strength, and near-average pressures over the United States. Temperatures were moderately low but near the twentieth-century means in 1761–1770. In successive decades temperatures generally rose in the West and declined in the East. Precipitation was above average in the North and below average in the South.

The driest decade in this interval was coincidental with severe drought reconstructed in the eastern United States in 1769–1776 by Cook et al. (in press). Maps of the seasonal reconstructions using the DIFMAP program (Appendix 3) show dry winters and springs in the southern Plains and dry but cool summers in the northern and central Plains and Great Lakes. These results agree with some of the Cook et al. data, which show dry conditions in two western regions of their grid. Drought reconstructed for two eastern regions was not confirmed, but these regions were to the east and well outside the verifiable area of the western-based dendroclimatic reconstructions.

7. 1791–1820

The only major volcanic eruption in the interval of 1791 to 1820 was Tambora in 1815; it was a low-latitude eruption with a large DVI of 3000. It is notable that the warming in this interval is associated with only moderate volcanic activity during the first 25 years (Lamb 1970), although Lamb had inferred higher dust levels than those observed because of the low temperatures reported during that time period. The differences in reconstructed sea-level pressure, temperature, and precipitation were again the reverse of the previous difference map. The directions of the changes were similar but somewhat larger than in the 1717 map, and the values were more often significant. The reconstructed temperature rise that was most marked at the beginning (Figs. 7.1 and 7.2) was surpassed only by the twentieth-century warming. Pressures declined as much as 2 mb over the North American Arctic and the North Pacific. Rising pressure anomalies over the western and central United States may have generated a weak pressure block. The Aleutian Low appeared to intensify, and more storms may have traveled over the northern portions of the continent. With lower pressures in the Arctic, there may have been fewer cold air outbreaks; prevailing airflow was from the south; temperatures increased everywhere on the map. With an enhanced northward flow of moisture from the subtropics, precipitation increased in the extreme Southwest. Precipitation in southern California was reconstructed to have increased by 10 percent. The reconstructed precipitation was still below the twentieth-century average values for southern California, so this change amounted to only an abatement of drought. Fewer storms may have moved through the Pacific Northwest, and drought appears to have been more common from the Great Basin to the Canadian prairie provinces.

In 1791–1800 (Fig. 7.12) an anomalous high-pressure ridge may have extended from the Klondike to the United States Plains. A low-pressure anomaly was reconstructed in the central North Pacific, as well as in the top two corners of the map. Warm temperatures in the West and cool temperatures in the East suggest that a strong meridional flow was present. Moderately low precipitation was reconstructed from California to the Great Lakes.

The pressure block strengthened over the United States in the 1800s and 1810s (Fig. 7.13), but the Aleutian Low strengthened and was centered farther north, and more storms may have entered the Canadian Arctic (also see Lough in press). Cold air outbreaks from the Arctic may have been greatly reduced, with anomalous warmth reconstructed for most of the map. Storms may have entered North America in Canada and perhaps in the extreme Northwest, bringing some moisture along the Canadian border and to the eastern United States. Low precipitation was reconstructed to have prevailed in the Plains and Southwest.

Cook et al. (in press) report that four of the driest pentads they reconstructed for the eastern United States fell within the 9-year period spanning 1814 to 1822. The annual maps for this interval do not confirm a drought, because autumns were reconstructed to be wet. Precipitation was reconstructed to be below the twentieth-century means from California to the Great Lakes for winter, spring, and summer, and tempera-

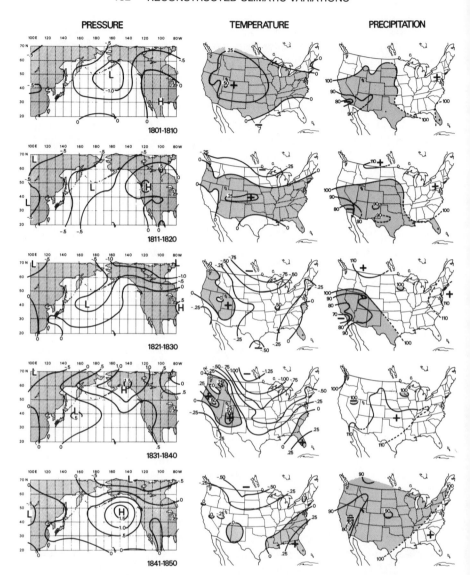

Fig. 7.13. Same as Figure 7.9, for the first five decades of the nineteenth century.

tures were very high, with many significant estimates for 1814–1822. Thus, there is considerable agreement between the two climatic estimates in the area of overlap if we consider the climate of the spring and summer seasons. But there are also limitations in comparing annual estimates obtained from conifers in the West responding directly to winter moisture and inversely to temperature with estimates of summer drought obtained by calibrating deciduous forest species with PDSI correlated with water stress. Another feature of this comparison should be noted. The autumn estimates were the poorest of the four seasons, and it would not be surprising for them to be least consistent with the independent estimates even if they had been for the same region. The lack of agreement of climatic reconstructions in the West with extremely cool conditions in the East is described in more detail by Lough (in press).

8. 1821–1849

Three volcanic eruptions of merit were associated with the 1821–1849 period: Galunggung (1822), Cosiguina (1835), and Hekla (1845). The first two, along with Tambora (1815), were from low latitudes; and Cosiguina had the highest DVI—4000—which was the highest in the 1602–1900 record. Sea-level pressures were reconstructed to have risen throughout Asia, the North Pacific, and the United States, but the changes were significant mostly in the eastern North Pacific at longitudes W 130–170° and latitudes N 30–50°. Decreasing sea-level pressures were reconstructed from the Alaskan Arctic to the United States northern Rocky Mountains and Hudson Bay. Although these differences were not significant, they suggest that there may have been a greater tendency for troughs to develop over western Canada. The strength of the Aleutian Low may have decreased while the North Pacific High increased. This pattern suggests that fewer storms may have traveled across the eastern North Pacific, reducing the frequency of storms entering along the Northwest coast. The inferred counterclockwise flow from the anomalous high and the clockwise flow from the low would be consistent with cooling in the United States.

The mean temperature change was the second most important change in the reconstructed record. Temperatures declined markedly throughout most of the grid. Differences were as low as $-1.06°$ C and suggest, along with the pressure grid, that cold air flowed southward, bringing cool weather, especially east of the Rocky Mountain Front. Precipitation was reconstructed to have been higher in some localities, perhaps because of the pressure trough in the West. From central California to the Canadian southwest, over the Great Lakes, and over many eastern states, precipitation was reconstructed to have declined several percentage points. No reconstructed precipitation difference over the 29-year period was significant, however.

In 1821–1830 an anomalous low extended from the western North Pacific northward across the Canadian Arctic and may have joined with the Icelandic Low to the east. Only a small area in the West showed warm dry anomalies, and the decade's average temperature and precipitation patterns were not greatly different from the embedded anomalies of 1602–1900 (Fig. 7.6). Cool and moist conditions were reconstructed from the Pacific Northwest to the eastern two-thirds of the United States. The agreement of the temperature estimates with the Robock (1979) simulations using volcanic activity was described earlier.

The decade of 1831–1840 was the wettest reconstructed since 1600 over the area covered by these maps, and the northern Plains were as cold as in any other decade reconstructed over the 1602–1962 interval. Sea-level pressures were reconstructed to have been anomalously high for Alaska and northwestern Canada, and also to the southeast over the United States. In the top two corners of the grid, sea-level pressure was reconstructed to have been low. Temperatures were warm or near the twentieth-century averages in the Far West, while they were reconstructed to have been severely depressed in the interior to the Atlantic Coast. The strengthened Klondike High, and possibly a strengthened Icelandic Low, could have increased the southward flow of cold air into the central and eastern United States. With a weak and perhaps southwesterly displaced Aleutian low, storms may have taken more southerly tracks across the North Pacific, entering North America south of the normal position along the California coast.

It is remarkable that the 1827–1837 interval was one of the wettest reconstructed by Cook and Mayes (1985) and Cook et al. (in press) in the eastern United States. When maps of seasonal temperature and precipitation reconstructions are constructed for this interval, the annual average precipitation exceeded twentieth-century levels for all areas of the United States except the southwestern coast and the Great Basin. The precipitation anomaly was especially marked for summer and spring, and summer temperatures were significantly below twentieth-century values for 39 percent and 22 percent of the grid points.

The sequences of possible dust veils from Tambora, Galunggung, and Cosiguina, all at low latitudes, associated with large-scale cooling and wet conditions in the Great Plains and eastern United States, support the possibility that long-term temperature and moisture conditions over this grid could be associated with low-latitude volcanic activity and related dust veils at lags greater than 3 years (Lamb 1970; Budyko 1969, 1974). Self et al. (1981) present convincing evidence that the eruption of Cosiguina in 1835 produced only a moderate volume of ash, and they claim that the low-frequency response of temperature during this time period should not be attributed to volcanic activity. This evidence, along with the findings of Kelley and Sear (1984), Bradley (1988), and Lough and Fritts (1987), casts doubt on the possibility that such long-term lagging relationships exist. The large contrast of this cool and moist period with drier and warmer conditions early in the century, confirmed to have existed in the East by Cook et al. (in press) as well as in the West, raises the question as to what climatic forcing function other than volcanic activity would account for such a large anomaly in climate.

In 1841–1850 an anomalous high was reconstructed south of the Gulf of Alaska in the eastern North Pacific, and an anomalous low stretched from eastern Siberia to Hudson Bay. This pattern appeared to advect more cool air into the West, but temperatures rose elsewhere in the United States. Blocking situations that may have been common along the North Pacific Coast could have reduced the number of storms and led to dry conditions throughout most of the United States. This decade was the beginning of an era of drought throughout the country.

9. 1850–1876

Two volcanic eruptions can be noted in the interval of 1850 to 1876, Sheveluch in 1854 and Askja in 1875. Both were at high rather than low latitudes and do not show a statistical relationship to short-term climatic variations (Lough and Fritts 1987). The changes in sea-level pressures for this interval were reconstructed to have been small and insignificant. The maximum rise was 0.8 mb at 80°W and 70°N, where an anomalous trough existed in the previous difference map. The Aleutian Low may have shifted somewhat to the south, and the pattern that advected cold air from the north was less prominent or absent. Temperatures were reconstructed to rise in the areas where they had declined in the previous map, and many differences were significant. Precipitation appeared to decline significantly at many grid points, however, especially in the central Plains, where differences were as low as −15 percent (see Figs. 7.5 and 7.8).

In 1851–1860 (Fig. 7.14) the anomalous high sea-level pressure that was reconstructed in the eastern North Pacific in the previous decade persisted, but the anomalous trough in the North American Arctic was not present. The southward flow of cool air was reduced, but blocking continued. The West warmed to the 1602–1900 embedded pattern, but precipitation was reconstructed to have been below the twentieth-century values throughout most of the United States. No large anomaly was noted by Cook et al. (in press) for this period.

In 1861–1870 the high-pressure anomaly disappeared, and the Aleutian Low appeared to strengthen and extend to the west, while sea-level pressures were reconstructed to have been higher than twentieth-century values over southern Canada and the United States. This pattern may

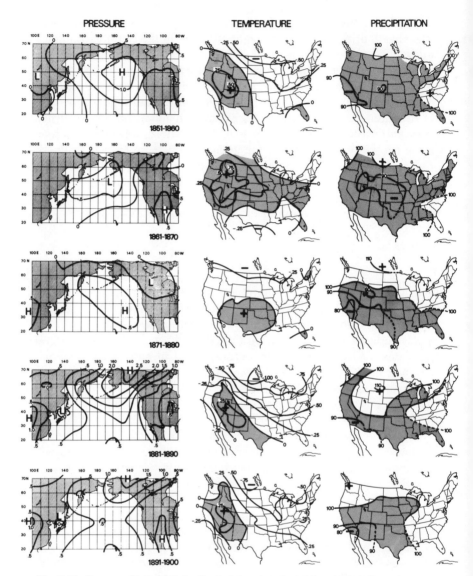

Fig. 7.14. Same as Figure 7.9, for the last five decades of the nineteenth century.

have favored a northward flow of warm, dry air and strengthened the high-pressure block. The pressure anomaly exceeded 0.5 mb in Mexico and the adjacent United States. Many cyclonic storms may have moved from the area of the Aleutian Low, along a northerly track across Alaska to the Arctic coast. Probably few storms moved to the south, adding to the severity and length of the United States drought.

In 1871–1880 the sea-level pressure anomalies were very weak, with positive values in the North Pacific and negative values largely confined to Canada, the northwestern United States, and the Alaskan Arctic. The temperature and precipitation anomalies were also reconstructed to have been weak with a cool, moist North and a warm, dry South and Southwest. The lower pressures in western Canada may have favored storm movement through that area with decreased storm activity in the southern United States. This decade marked the end of the drought in the North, but dry conditions persisted throughout the South.

10. 1877–1917

Two eruptions are of merit for the portion of this interval in the nineteenth century. Krakatau (1883) was the most notable, with a DVI of 1000, and was a low-latitude source. Bandai (1888) was a midlatitude source. Sea-level pressure increased over large areas of the map. The increase was as much as 1 to 2 mb in the North American Arctic and eastern Asia. Significant increases in sea-level pressure were reconstructed throughout the western North Pacific. Decreased pressures were reconstructed only for Washington and Oregon, reducing any blocking tendency there. The difference patterns resembled the 1761 differences with reduced strength of the Aleutian Low and possible expansion of the Mackenzie and North Pacific Highs. The winter storms in the eastern North Pacific may have traveled on more southerly routes entering North America along the United States Pacific Coast. Temperatures declined as much as 0.66° C in the northern Plains and appeared to have been lower throughout all areas but the southern border of the grid. Precipitation was reconstructed to have increased except for scattered areas in the extreme South and Southwest.

In both 1881–1890 and 1891–1900 (Fig. 7.14) a strong high-pressure anomaly was centered over the Yukon and extended southwest over the North Pacific. The circulation may have been strong, bringing severe cold and moisture to the northern and eastern United States, but the milder embedded anomaly is evident throughout the Southwest. The reconstructed temperature patterns were the same as before, although the cooling along the United States-Canadian border was less. Dry conditions, which were reconstructed in the Southwest, extended into the Plains states in the 1891–1900 map. The 1883 eruption of Krakatau was associated with cooling in the central and eastern states with an increase in precipitation 1–3 years following the eruption, but it was the only low-latitude eruption and had no apparent prolonged effect on climate.

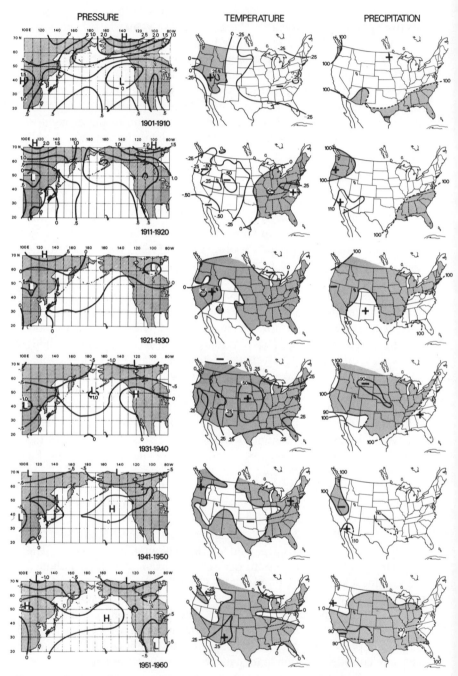

Fig. 7.15. Same as Figure 7.9, for the first six decades of the twentieth century.

The remaining decade maps (Fig. 7.15) are the reconstructions for the dependent period. They were forced by the calibration to match the instrumental record, and since they are in the twentieth century, the reconstructed anomalies from the twentieth-century mean values are much reduced. They are included here with the independent reconstructions only to relate the known twentieth-century changes to the features reconstructed in this work. The reconstructions were forced to match the instrumental data by the calibration, so there is no need to examine other data for validation. The seasonal climatic anomaly types of Blasing and Lofgren (1980) were used to characterize the seasonal climatic patterns that dominated each decade map.

The anomaly for sea-level pressure in 1901–1910 was similar to the anomaly reconstructed in 1881–1890, except for one important difference. The anomalous high in the 1901–1910 period did not extend as far south, and a weak anomalous low was embedded in the pattern over the North Pacific off the Washington and Oregon coasts. This pattern resembled the sea-level pressure patterns reconstructed in 1741–1750, except that the low was closer to the North American coast. Temperatures were substantially warmer in the East as cold air outbreaks from the north were probably less severe or less frequent. Temperatures in the Southwest were below those of the embedded 1602–1900 period. Precipitation was above the twentieth-century average figures for large areas of the map. These patterns suggest that more storms may have entered the northwestern United States than before, bringing moderate temperatures to the Great Basin and the Northwest and cooling along the coast with more precipitation over most of the United States. The circulation pattern appeared more zonal in 1901–1910 than before, with contrasting pressure anomalies in different quarters of the map. The overestimation of Arctic pressures in the Dawson region before 1910 (Trenberth and Paolino 1980; Jones 1987) may have contributed to some errors in the pressure reconstructions for the Far North.

Blasing and Lofgren (1980) show that the 1900 instrumental records of sea-level pressure were dominated by high-pressure anomalies in the Artic for all four seasons. The Aleutian Low was strengthened and easterly displaced in summer and autumn, resembling the 1901–1910 reconstructed pattern. The temperatures varied during the seasons but were generally cooler than the twentieth-century mean values. Autumn type 4 was the only pattern with warming in the Northwest. Precipitation was also variable among the four seasons. The only area of average-to-low precipitation for all four seasonal types was in the Southeast. The typed patterns were consistent with the annual decade estimates.

Similar sea-level pressure anomalies were reconstructed in 1911–1920, with smaller anomalies in the North American Arctic and larger anomalies in the Asian Arctic. Pressures were somewhat higher along the United States North Pacific Coast and western Canada. The closest analog to this pattern was 1602–1610, but there was clearly no decade with a good match. These higher pressures may be associated with cooling in the West. The 10-year averaged temperature reconstructed for the 77-grid in the West (Table 7.1) was lower in 1908–1917 than at any other 10-year period since 1602. The average temperature reconstructed for the 46-grid in the West for 1909–1918 was fifth-ranking. Temperatures continued to rise in the East, however. Precipitation in the West (Table 7.2) for 1907–1916 was reconstructed to be the second highest in the 52-grid since the beginning of the reconstructed record. Precipitation was high and drought infrequent in the East during this period (Cook et al. in press).

The dominant pressure types of Blasing and Lofgren (1980) for this decade include a high anomaly in the Siberian Arctic, a high in the North American Arctic in autumn, and a strong Aleutian Low in winter. The temperature patterns were mixed. Autumn was the only type with clearly low temperatures in the West. However, the low temperatures and high precipitation during this time period are evident in the plots of the instrumental data shown in Figures 7.1 and 7.2 (Boden et al. 1990).

11. 1918–1960

The changes in these maps were dominated by rising temperatures and declining sea-level pressures and precipitation over large areas of the maps. The greatest pressure changes were −2 mb over the western North American Arctic. Pressure increased at only one grid point over the Pacific Northwest. This change suggests that the Arctic Highs became less important with increasing frequencies of storms moving across the North Pacific and entering the mainland through Alaska and the Canadian Arctic. The pressure gradient was reduced at midlatitudes, affecting the strength of the westerlies (van Loon and Williams 1980). Precipitation declined as much as 10 percent throughout the northern two-thirds of the grid. The South and Southwest were wetter, particularly in southern California, where the maximum rise in precipitation was reconstructed to have been over 11 percent. These data imply a second area of increasing storm activity through California, Arizona, Colorado, and the South. The subtropical highs may well have expanded farther north, perhaps bringing more tropical moisture into the Southwest and more warm air into the central United States to the

Atlantic Coast. The average position of the Arctic Front may have retreated to the north.

The only notable anomalies in 1921–1930 for sea-level pressure were a weak low over central Canada and a weak high and low in western Asia. Warm anomalies were reconstructed everywhere but in the central Rockies, the Southwest, and the northwest Pacific Coast. Precipitation was reconstructed to have been near the 1901–1970 average with a slight dry anomaly everywhere but in the southwestern deserts. The pressure types of Blasing and Lofgren (1980) show a strong Aleutian Low and a strong Klondike High with no marked temperature or precipitation trends evident. Figures 7.1 and 7.2 confirm the trends in the instrumental record.

In 1931–1940 the Aleutian Low strengthened and sea-level pressure over the United States was higher than the 1901–1970 average, suggesting that blocking situations may have increased. The early warming reached its peak, especially in the East (Table 7.1), and drought (Table 7.2) was notable over the interior, but the 10-year mean value for precipitation was not as low as it was for some pre-twentieth-century reconstructed maps. The dominant pressure types of Blasing and Lofgren included a strong Aleutian Low in spring, and a strong North Pacific High in summer. Both conditions were associated with drought in the western part of the United States. However, the dominant winter type contradicted this pattern.

In 1941–1950 sea-level pressures were anomalously low in the Arctic and slightly above the twentieth-century averages in the eastern North Pacific. The blocking tendency was reduced, and the midlatitude storms may have traveled more northerly tracks. The subtropical jet may have become more important in advecting moisture and mild temperatures into the Southwest. Above-average moisture was reconstructed for most of the United States, and it was cooler, especially in the midcontinent. The dominant types of Blasing and Lofgren show a strong Aleutian Low, low pressure in the North American subarctic, and a strengthened North Pacific Subtropical High, particularly in the eastern North Pacific. The typed temperature and precipitation conditions are variable and difficult to relate to the reconstructions, but the instrumental data in Figures 7.1, 7.2, and 7.4 follow the general patterns of the reconstructions.

It was somewhat warmer and drier in 1951–1960, according to these maps. Low-pressure anomalies in the Arctic suggest that the northerly storm tracks remained important. However, the Aleutian Low was not as strong as it was in the 1930s, a development that would be consistent with fewer or weaker storms entering along the California coast and less precipitation in all but the northern United States. All selected types have anomalously high sea-level pressure in the eastern North Pacific. Some areas were typed as cool in the West. Dry conditions were typed in the spring and summer, particularly in the South, and the winters were typed as more moist.

Although the sea-level pressure anomalies in the annual reconstructions frequently do not coincide with the detail evident in the seasonal pressure types from the twentieth-century data (Blasing and Lofgren 1980, the relationships of the pressure types to seasonal temperatures and precipitation are similar to the relationships observed in the seventeenth through nineteenth centuries among the annual sea-level pressure, temperature, and precipitation. Typing the annual reconstructions did not produce a useful system of classification.

8

Discussion

It is possible to identify some likely causes of the reconstructed climatic variations presented in this book, but the largest part of the variance is not clearly related to any single causal factor. Dust-veil forcing of climatic variation from explosive volcanic eruption at lags of up to 2 years' duration is clearly an important factor. Volcanic activity relationships for longer lags are suggested, but the reality of such a relationship has not been demonstrated to be statistically significant. Some variance can be attributed to ENSO phenomena such as the SO (Lough and Fritts 1987, 1990). Some variance may also be associated with sea-surface temperatures and sea-ice variations, but the past history of these phenomena has not been worked out in sufficient detail for the area affecting western North America to be used for this analysis. Possible relationships to solar variability appear to influence some variance, but we did not repeat the rigorous analysis described by Stockton et al. (1983), which would be needed to answer this question. Though a large part of the variations could not be attributed to specific causal factors, other data sets were examined for similar or dissimilar variability.

One type of validity test asks these questions: Do the reconstructions behave like the instrumental data? Are the relationships among the reconstructed temperature, precipitation, and sea-level pressure the same as those among the temperature, precipitation, and pressure measurements? Are the relationships for the reconstructions in the calibration interval the same as those for the reconstructions in the earlier independent period?

Additional validation can be obtained if the relationships found in the reconstructions are consistent among (1) the three variables of temperature, precipitation, and sea-level pressure, (2), with the instrumental records of the variables, and (3) with other proxy evidence. This chapter first describes some applications of these data to problems of climatic variation and change. It then addresses some of these validity questions and considers the evidence for climatic changes related to Little Ice Age climates.

A. Applications

Blasing and Fritts (1975) examined tree-ring data from forests that border the American Arctic and compared growth to summer sea-level pressure anomalies reconstructed from the 65-chronology data set. They noted ten independent ring-width anomaly patterns between 1800 and 1939 and interpreted the changes in terms of the general circulation suggested by the reconstructed sea-level pressure maps. It was too early in the study to draw firm conclusions, but the systematic changes that were noted suggested that the reconstructed large-scale synoptic patterns in sea-level pressure were real features that could lead to a better understanding of past climatic variability and forest-stand changes. They recommended that more proxy data be assembled from the subarctic using modern techniques (Baron 1980) and that all such evidence on past climatic variation be interpreted jointly. They also suggested that it might be possible to improve the estimates of sea-level pressure by incorporating tree growth from northern areas into the transfer function. Guiot (1987) followed through with just such an analysis. His study was located outside the area of our temperature grid and at the margin of the pressure grid, however, and he did not use the pressure reconstructions or the arid-site tree-ring information. These studies did indicate that a year-to-year and possibly a season-to-season description of past paleoclimatic conditions in the Arctic was possible using diverse data sets. If the long-distance relationships evident in the pressure reconstructions could have been included with more local studies like those of Guiot, it might have been possible to span the distances and to deal with patterns over a wider area of the Arctic. When the results of the two independent studies were compared visually, agreement was not apparent.

Early descriptions of the methodology described in this book appear in Fritts, Lofgren, and Gordon (1979, 1980, 1981), but the first well-verified reconstructions were reported in 1981 (Fritts 1981), along with some of the improvements due to averaging. Some general principles suggested by these results were described by Fritts (1982, 1987) and were applied to problems of water scarcity and drought in the West (Fritts 1983, 1984).

Cropper and Fritts (1985) analyzed the adequacy of the twentieth-century climatic record as baseline data for managing a proposed nuclear waste disposal site in the Pasco Basin, Washington. They attempted to improve the estimates at three grid points surrounding the Pasco Basin which were made in this volume, by recalibrating and verifying the instrumental measurements with newly developed chronologies from the area provided by Brubaker and Graumlich (personal

Table 8.1. Differences between the Means and Ratio of Standard Deviations for the 1901–1970 Instrumental Record and the 1602–1900 Reconstructions of Temperature and Precipitation at Selected Stations in Washington State (Cropper and Fritts 1985)

Season	Temperature[a] (in °C)		Precipitation[a] (in mm)	
	Mean	Standard Deviation	Mean	Standard Deviation
Annual	− .096	0.962	− 8.153	0.777
Winter	− .344	1.076	− 1.702	0.808
Spring	− .179	0.814	3.810	0.926
Summer	0.029	0.872	2.870	1.090
Autumn	0.180	0.727	− 13.132	0.621

[a]The average of temperature measurements at Baker, Walla Walla, and Spokane and the average of precipitation measurements at Lewiston, Baker, and Collville. The reconstructions for the early period were subtracted or were the divisors.

communication 1986), Graumlich and Brubaker (1986), and others. The statistics from the new reconstructions were compared to the statistics from the older reconstructions for the same data points. Though the new reconstructions had the highest calibrated variance, the verification statistics were substantially lower than those for the earlier work, indicating that they were not as reliable. The new models were abandoned, and the analysis was completed using the original reconstructions for the four grid points.

The reconstructions for 1602–1900 were compared to the 1901–1970 instrumental record, and the statistics of the latter were adjusted for the longer reconstructions (Table 8.1). The averages and the standard deviations for both annual temperature and precipitation were too low for the twentieth century (Table 8.1), so they were adjusted upward. Average winter and spring temperatures were below those reconstructed in the seventeenth through nineteenth centuries, but summer and autumn temperatures were higher. The standard deviations for 1901–1970 were below those reconstructed for the seventeenth through nineteenth centuries, except for winter temperature and summer precipitation. The most pronounced difference between these two time periods was the higher variability reconstructed for the seventeenth through nineteenth centuries.

Fritts and Lough (1985) compared the large-scale features and low-frequency variations in the annual temperature reconstructions from

North America to the averages of instrumental records of Jones et al. (1982), Manley (1974), and Groveman and Landsberg (1979), as well as to upper tree-line and Arctic dendrochronologies. They noted that the twentieth-century warming was a prominent feature of all reconstructed temperatures. The annual temperatures were inversely related to annual sea-level pressures with a correlation of -0.62 over the entire length of the two records. The correlation was strongest between half-hemispheric averages of reconstructed pressure and temperature in the eastern United States, which is compatible to the findings of Diaz and Namias (1983) using twentieth-century data.

Fritts and Lough (1985) pointed out that large differences in temperatures have been observed for different regions around the world, such as those cited by van Loon and Williams (1976) and Kelly et al. (1982). Since the average reconstructed temperatures over the entire 77-station grid provided new information for a long time period from an important geographic area including both western and eastern North America, Fritts and Lough proposed that the time series be compared to other measures of geographically averaged temperature variations. When comparing the temperature reconstructions to the hemispheric average temperatures of Jones et al. (1982), Fritts and Lough found a high correlation for 1881–1946, but the correlation declined after 1946 when hemispheric temperatures and western North American temperatures were in opposition. The authors also compared the series with Manley's central England temperatures (Manley 1974) and found them significantly and inversely related around the middle of the eighteenth century but not for any other period. Since this result was not much higher than would be expected from chance variation, Fritts and Lough concluded that temperatures in the western United States are not synchronous with the Manley temperature record and decided that it would be unprofitable to attempt an evaluation of the North American reconstructions using European temperature records.

Next Fritts and Lough (1985) compared their data to the hemispheric temperature estimates of Groveman and Landsberg (1979) made from proxy records at varying locations and of different quality and length. The correlations with this series were positive and significant only during 1656–1680 and 1915–1940. As one might expect from the lack of synchrony with European temperatures, disagreement was most marked when Groveman and Landsberg's proxy records came largely from the European and North Atlantic sectors of the hemisphere and not from the United States. This result suggested that Groveman and Landsberg's use of different proxy records of varying quality for different time periods introduced large variations in the estimates that appear unrelated to real differences in hemispheric-wide temperature changes. Fritts and Lough noted that Groveman and Landsberg had used only a few tree-ring chronologies and that these were from the subarctic and of dubious quality. They proposed that the Groveman and Landsberg estimates for North American temperatures could be improved substantially by using the spatially averaged reconstructions from the arid-site chronologies along with the temperature estimates from well-dated high-altitude and Arctic chronologies described by Stockton et al. (1985).

Long and high-quality climatic estimates of this kind, which have been averaged over geographic regions, will be needed to evaluate model simulations of spatial variations from global average values. Long reconstructions from two potentially nonsynchronous areas such as western North America and Europe will be useful for evaluating modeled differences as they evolve through time and for evaluating the consequences of diverse climatic impacts on different economic systems.

Lough and Fritts (1985, 1990) have examined the high-frequency relationships between spatial temperature and precipitation variations in the United States and southwestern Canada and the SO in the South Pacific. They found similar teleconnection patterns with the SO in (1) the twentieth-century instrumental records, (2) the twentieth-century reconstructions, and (3) the reconstructions for the nineteenth century. This consistency in response was additional confirmation that the temperature and precipitation reconstructions have preserved the integrity of the atmospheric system. When the high-frequency SO features were related to the low-frequency reconstructions of sea-level pressure, however, the teleconnection patterns were not statistically significant. This result is consistent with the analysis of variance showing that sea-level pressure reconstructions were less reliable at high frequencies than the reconstructions of either temperature or precipitation (Chap. 6, Sec. F; Fig. 6.22).

Since a relationship with the temperature and precipitation reconstructions could be demonstrated, seasonal SO data were calibrated directly with tree-ring data from the Northern Hemisphere and from the Southern Hemisphere (Fritts and Shatz 1975; LaMarche et al. 1979a, 1979b, 1979c) using canonical regression. The calibration and verification statistics for the Southern Hemisphere data were lower than those for the Northern Hemisphere data. The two reconstructions contained only a small amount of common information about the SO, and the information that could not be calibrated, which is shown in the residuals, was very similar. The combination of both Northern and Southern Hemisphere tree-ring chronologies did not provide any significant improvement, so only the North American chronologies were used for the final calibration and verification. The estimates explained about 50 percent of the SO variance and 20 percent of the variance verified over the independent period. Significant high-frequency spectral peaks were noted at periods from 2 to 10 years like those found in the record of the SO. Low-index events were reported to have been less frequent during the nineteenth century than before or since.

Lough and Fritts (1990) point out that the tree-ring record from temperate climates cannot provide the important information of ENSO relationships in tropical regions. Until dendrochronologies from tropical regions can be developed (Fritts and Swetnam 1989), other proxy records from the tropics such as coral skeleton variations (Lough and Barnes 1990), ice cores (Thompson et al. 1984, 1985, 1986), and varves (Baumgartner et al. 1989) may provide the missing information on tropical ENSO patterns. Many of the techniques described in this book, by Guiot (1987), Cook et al. (in press), and Fritts et al. (1990), are well adapted to the analysis of tree-ring data and other proxy information with diverse responses to climatic variations.

The effects of volcanic eruptions on twentieth-century climate are limited by the small number of eruptions that have occurred during the century. More volcanic activity was recorded in the nineteenth century, but the coverage of instrumental climatic records for that time period is inadequate. Lough and Fritts (1987) substituted the dendroclimatic reconstructions for the early instrumental records and used "superposed epoch" analysis to examine the impact of past explosive volcanic eruptions on spatial climatic variations. Monte Carlo techniques were applied to assess the statistical significance of the average result. No significant effects were detected from eruptions originating between 20 and 50°N, but highly significant effects were obtained for eruptions at latitudes between 10°S and 20°N. The response was not uniform over the grid. Cooling was observed in the central and eastern United States during spring and summer for the first 5 years following the eruption events. Warming was observed in the western half of the United States in winter and in the Far West in summer.

Independent tree-ring chronologies from high-altitude sites with a direct temperature response, along with other proxy records of climate, were used to validate these observations. Five Rocky Mountain chronologies that were in the area of marked cooling also had a reduced growth following large volcanic eruptions. This growth reduction confirmed that cooling occurred during the growing period in the Rocky Mountain states. Five Great Basin chronologies that were in an area of winter warming and summer cooling also showed reduced growth following eruption events, but the average values of the growth departures were not significant. The effect of summer coolness may have been offset by the winter warmth so that the average volcanic response was less marked and insignificant. Temperature reconstructions from high-altitude trees for Longmire, Washington (Graumlich and Brubaker 1986), weakly supported the conclusion that warming occurred in the Far West, especially during the second year after low-latitude volcanic eruptions. Although these tree-ring data provide inconclusive evidence for warming, they did confirm the absence of cooling in the Far West following low-latitude eruptions.

Cooling in the eastern United States was confirmed by the Philadelphia temperature record of Landsberg et al. (1968). Furthermore, three proxy climatic series from polar regions suggested that temperatures north of the area of valid tree-ring reconstructions also may have declined after low-latitude volcanic eruptions. The Lough and Fritts (1987) study points out the usefulness of continuous reconstructions of seasonal climatic information before the existence of instrumental climatic data.

The short-term climatic response to volcanic dust veils reported by Rampino and Self (1982), Kelley and Sear (1984), and Bradley (1988) is not only confirmed in the reconstructions for the seventeenth through nineteenth centuries, but a warming response is identified in western North America where the tree-ring reconstructions appear to be best. Cooling, sometimes associated with high precipitation, appears to have followed some large eruptions at lags greater than 6 years (Chap. 7; Lamb 1970). However, the significance of such a long-lasting effect may require more proxy information from earlier time periods to provide adequate replication for testing.

B. Comparisons with Other Tree-Ring and Proxy Data

Comparisons with other proxy evidence can be made at two distinct levels. First, different data sets from the same climatic region can be compared to evaluate differences in proxy response and errors in the estimates, as well as similarities in the reconstructions. Second, large-scale features reconstructed for different regions also can be compared to known large-scale features of the climatic system. Some features in this second category already have been described by Blasing and Fritts

(1975), Gordon et al. (1985), Fritts and Lough (1985), and Lough and Fritts (1985, 1987, and 1990).

The first level of comparison may involve two kinds of estimates: (a) discontinuous measurements, often with approximate dating, and (b) continuous measurement with precisely dated, well-behaved, annual time series. Most glacial evidence and historical accounts of climatic events fall into category (a). These data often come in bits and pieces that are hard to compare to dendroclimatic reconstructions because of the uncertainty regarding how accurately the information is dated, what climatic factor is recorded, and what time frame (day, week, month, season, or group of seasons) is involved. For example, the frost-ring events of LaMarche and Hirschboeck (1984) and their relationship to climatic fluctuations, dust veils, and volcanic eruptions are not easily compared to reconstructions of summer temperature, because the frost rings involve damage to a few trees, usually from a one-night event, whereas the reconstructions involve a response from many trees over the growing season. Observations on glacial advances and retreats are not as well dated as tree-ring chronologies; lags of many years may be involved, and the exact climatic variable controlling the variation may be less certain (Bradley 1985) than the variables calibrated and verified to obtain dendroclimatic reconstructions.

Some of the most valuable paleoclimatic information comes from category (b), and often this information is based directly or indirectly on some form of tree-ring analysis, including measurements of isotopic content of individual tree rings (Stuiver 1978, 1980; Stuiver and Quay 1980). Isotopic analysis is customarily made from samples spanning more than one annual ring (Leavitt and Long 1988), so little information exists on yearly and spatial variations. In addition, the variability of isotopic measurements within and between trees has not been fully assessed (Wigley et al. 1978; Jacoby 1980). The availability of tree-ring chronologies and other annual data from western North America is discussed in Chapters 1 and 2.

The large-scale temperature reconstructions described in this book were compared to other independent proxy records (Fritts and Lough 1985), including the high-altitude chronologies of LaMarche and Stockton (1974). The temperature reconstructions (Fig. 7.1) did not indicate as marked a rise in temperature during the late nineteenth and the twentieth century as had been inferred from some high-altitude tree-ring chronologies from western North America by LaMarche et al. (1984). The trends in two out of the four tree-line time series that had been considered were significantly correlated ($p \geq 0.95$) with the temperature estimates over the 1901–1960 period. The correlations of the regionally averaged temperature series with these two high-altitude tree-ring series were no longer significant when the longest period of overlap was correlated, however.

1. The High-Altitude Response

Graumlich and Brubaker (1986) relate ring-width indices of *Tsuga mertensiana* and *Larix lyallii*, growing near the high-altitude tree line in the Cascade Range in Washington, to temperature, precipitation, and snow depth. To investigate the possibility for interactions among variables, the authors examined only the variables that had the highest apparent correlation with ring-width variation. These included July-to-September temperatures and precipitation and snow depth for January and March. Graumlich and Brubaker report that when snow depths were average or less, both species showed simple patterns of increasing growth with increasing summer warmth. At higher snow depths there was no temperature-growth relationship, or the relationship with temperature was inverse. The authors note that November-to-March temperature was significantly negatively correlated with March snow depth, and they argue that the relationship between growth and annual temperature would implicitly include some inverse effects on growth due to spring snow depth.

Graumlich and Brubaker (1986) do not examine the temperature and snow depth relationships further using either response functions or response-surface analysis. They do use calibration and verification procedures, including an identification of reconstructions that were extrapolations beyond the range of the calibrated record. They obtain annual temperature reconstructions that were corroborated by glacial evidence (Burbank 1981; Heikkinen 1984), but except for reconstructing the warming trend at the end of the nineteenth century, they note little agreement with the Jacoby et al. (1985) reconstructions of summer-degree days in northwestern Canada or with the Campito Mountain high-altitude *Pinus longaeva* chronology (LaMarche and Stockton 1974) from California. These differences may represent real differences in the temperature variations over space or may be attributable to differences in the tree-ring response to variables other than temperature.

Graumlich and Brubaker (1986) also compare their temperature reconstructions to those of Cropper and Fritts (1985), which include three stations in the Columbia Basin, shown in Figure 7.2. Graumlich and Brubaker are concerned that their reconstructions indicate cool conditions throughout most of the eighteenth and nineteenth centuries, followed by a 0.9° C rise in temperature beginning in the twentieth century, whereas the Cropper and Fritts reconstructions show little change in the mean temperature over the last 300 years. Graumlich and

Brubaker attribute the differences to the varying sensitivities of sub-alpine versus semiarid trees in response to long-term changes in temperature.

The temperature reconstructions involving trees in cold high-altitude or polar habitats show different patterns from site to site and do not necessarily agree with the temperature reconstructions described in this book from trees growing on arid sites. A number of possible causes exist for these differences. Tree-ring growth at both the upper tree line and the lower forest border can be limited by factors other than temperature, and the relative importance of temperature can vary greatly from one season to the next (Fritts 1974, 1976; Graybill, personal communication 1989). Interactions between different limiting conditions may be present, and curvilinear relationships can exist (Graumlich and Brubaker 1986; Graumlich 1991). For example, some high-altitude, upper tree-line trees can respond to both moisture changes and snow accumulation (Graumlich 1985; Graumlich and Brubaker 1987) or can respond selectively to temperatures in different seasons, and the response to temperature is not always direct (Fritts 1974). Inverse response to winter temperatures (Fritts 1974) may be attributable to high respiration in the absence of winter-time photosynthesis and depletion of food reserves.

Graybill (personal communication 1989) has examined the temperature relationships in his chronologies from high-altitude, upper tree-line Great Basin sites and cannot find a strong positive correlation greater than $+0.5$ with either monthly temperature or monthly precipitation. He acknowledges that this difficulty can be attributed to the lapse rate and other differences between the low altitudes of the climatic stations and the high altitudes of the tree sites, but his attempts to correct for the obvious differences did not yield improvements in the relationships.

Other differences exist between high-altitude or high-latitude chronologies and low-altitude chronologies from arid sites. The high-altitude chronologies exhibit more low-frequency information, which may require ARMA modeling before being interpreted as a record of long-term temperature trends (Graumlich and Brubaker 1986; Graybill 1987). The specific mechanisms that produce low-frequency information in one type of series but not in the other need to be identified, as well as differences in the seasonal temperature and moisture response. In addition, some high-altitude, upper tree-line chronologies exhibit anomalous trends in recent years that could be a CO_2 fertilization effect (LaMarche et al. 1984), but these trends cannot be found in chronologies from sites with moderate temperature or water-stress response (Kienast and Luxmore 1988) or from high-altitude and moist Sierra Nevada sites (Graumlich 1991). Graybill (1987) and Graybill and

Shiyatov (personal communication 1989) are examining other high-altitude, tree-line environments for a possible CO_2 growth response but so far have not found evidence for it in another area.

2. High-Altitude Trees
and the Glacier Response

Glacier fluctuations (Bradley 1985; Wood 1988a; Porter and Denton 1967) result from changes in the mass balance of glaciers involving the net accumulation of ice volume, mass transfer, ablation and melting, as well as advances or retreats at the glacier snout. Glacier advances and retreats generally originate from temperature variations during the melting season and snow accumulation. Both variables can vary over the season and from one locality to the next, and local features of the glacier can be important.

High-altitude trees often respond to the same variables of climate, so that the low-frequency variations in growth can correlate with glacial advances and retreats (LaMarche 1978; LaMarche and Fritts 1971; Graumlich 1985; Graumlich and Brubaker 1986; Scuderi 1987). Often these correlations are cited as evidence that annual temperatures are changing (LaMarche 1974; Lamb 1977; Bradley 1985), but the variations in the tree-growth response to temperatures of the different seasons, inverse relationships with temperature, and possible influences of precipitation are assumed to average out or to be unimportant.

3. The Low-Altitude, Arid-Site Tree Response

Precipitation is assumed to be the factor most likely to limit ring-width growth on temperate sites (Stockton et al. 1985). High temperatures can enhance evapotranspiration and water loss, however, and thus reduce growth (Fritts 1976). If the effects of temperature and precipitation can be combined into a water balance (Fritts, Vaganov et al. 1991) or a PDSI, there may be a substantial improvement in the correlation of ring widths with climate (Cook and Jacoby 1977, 1979; Cook et al. in press; Stahle et al. 1988; Stahle and Cleveland 1988; Stockton and Meko 1975, 1983). The calibration of ring-width variation with the PDSI can provide highly reliable reconstructions of meteorological drought but cannot be interpreted as easily as precipitation or temperature measurements.

Instead of combining temperature and precipitation to obtain a drought estimate, this book considers reconstructions of temperature, moisture (precipitation), and sea-level pressure as independent factors, so that spatial variations can be examined in the context of standard meteorological measurements. This is the first time that temperature variations have been reconstructed from arid-site tree-ring chronologies, along with precipitation, and the reliability compared and mapped (also see Villalba 1990). The features of these three variables are interpreted as patterns of airflow, possible frontal activity, storm tracks, blocking, and other deduced synoptic-scale features in the reconstructed maps.

A regression analysis of this kind cannot always discriminate among colinear relationships and identify exactly which features represent temperature and which represent precipitation estimates, however. Although objective tests have been applied in a carefully designed analysis to diagnose the feasibility of the methodology and to deduce joint patterns in climate, the result is a compromise and, at best, only a first approximation of the real-world system, as indicated by large error terms and residual variance.

The annual temperature reconstructions were the averages of the temperature responses in the four seasons, which reduced some of the error variance. A multivariate transfer function was used to calibrate each season separately to help sort out the inverse from the direct temperature relationships that were evident in response-function results (Fritts 1974). The attempted separation into reconstructions for each season was incomplete, because overestimates associated with extreme climate in one season were often accompanied by underestimates in other seasons (the climatic data for each season were calibrated with the ring widths that had responded to the integral of the yearly climate). However, this overestimation in one season and underestimation in another season amounted to compensating errors that often canceled out one another when the seasonal reconstructions were averaged to obtain the annual climate. The agreement between the annual instrumental measurements and the annual reconstructions substantially increased over the average of the seasonal results (Figs. 6.9–6.15 and 6.17–6.18). In addition, the seasonal estimates of temperature were the average of relationships for 65 chronologies, may have involved more than one lag in the response, and usually included two to three models that were averaged to obtain the optimum results. Since the annual temperature estimates included the temperature variations averaged over all four seasons, not just warm-season months, and were the average of many chronologies and a variety of models, they should be relatively free of large errors and a good estimate of the annual average temperature. The error reduction was at a cost, however, because it reduced some of the time resolution and high-frequency variations in climate.

Although the spatial reconstructions of temperature from arid-site trees turned out to be more reliable than spatial reconstructions of precipitation, they are probably less reliable than the estimates of the PDSI, which combines both temperature and precipitation but leaves out other temperature effects not related to the water balance. Temperature estimates from arid-site relationships are more or less independent of temperature evidence from high-altitude and high-latitude sites. Arid-site estimates of temperature use a different component of the energy budget and show both similarities and differences from other types of temperature evidence. Though the reliability of these new independent estimates of temperature variations should be scrutinized carefully, the fact that they are different from upper tree-line or high-latitude evidence for temperature variation does not automatically refute their credibility. It does suggest that all types of proxy evidence of past temperature variations throughout North America should be reexamined to identify possible causes of these differences, sources of error, and uncertainties in our knowledge of both long-term and short-term temperature changes.

C. The Little Ice Age

The last decline in tree line in the United States occurred before A.D. 1500 (LaMarche 1973), and warming occurred after that period. Lamb (1977) states that the climax of the Little Ice Age seems to have come at different times in different regions and was earlier over North America than over Europe. He notes that the last advance of glaciers, around 1780–1850, in contrast to earlier periods, was related more to an increase in precipitation than to a decrease in temperature. He characterizes the central period of the Little Ice Age, 1500–1700, as a time with (1) low values of the long-term mean temperatures in all seasons of the year, with some exceptions, and variations in timing at northern latitudes; (2) enhanced variability in temperature and on somewhat longer time scales with frequent blocking, or meridional circulation patterns, at middle latitudes; and (3) atmospheric circulation patterns associated with an expanded polar cap and circumpolar vortex. It is not easy to discern patterns of cooling and warming in these climatic reconstructions from arid-site dendrochronologies which are clearly related to glacial advances and retreats during the Little Ice Age (Grove 1988; Ladurie 1971; Lamb 1977; LaMarche 1974). Enhanced variability (character 2) and circulation features related to an expanded circumpolar vortex (character 3) were reconstructed, however.

Lamb (1969) examines the Northern Hemisphere evidence for anomalies in circulation during the Little Ice Age and finds an increase in wave number and a change in the position of upper-level troughs. Most of his evidence comes from the North Atlantic and European sectors, and he often assumes that the circulation was anchored on the North American Rocky Mountain crest. In maps for 1500–1600 Lamb locates troughs off the west coast of North America, but a strengthened trough almost always was indicated over eastern Canada with a southerly displaced Icelandic Low and strengthened southerly flow of cold air into eastern North America and the North Atlantic. Lamb (1969) also analyzes the average mb thickness for 1780–1820 compared to 1950–1958. He finds similar conditions to the 1500–1600 period but notes that the anomalies in the Pacific were probably smaller. The 1780–1820 anomalies are attributed to a small reduction in the average insolation reaching the surface of the globe associated in part with dust veils from volcanic activity.

During the seventeenth through nineteenth centuries the areas between the Cascade Range and the Sierra Nevada on the west and the Rocky Mountains on the east were reconstructed from arid-site tree-ring chronologies to have experienced average temperatures slightly above the 1901–1970 mean values. This anomalous warmth may be in part the result of the cool interval in the Far West in the 1910s and 1920s, when temperatures in many parts of the world may have been rising. East of the Rocky Mountain crest, however, temperatures were reconstructed to have been markedly cooler than twentieth-century values. This cooling is in agreement with Lamb's (1969) finding of enhanced southerly flow from the Icelandic Low into eastern North America and the North Atlantic. Associated with this apparent warmth in the West and coolness in the East, sea-level pressures over North America, especially in Alaska and western Canada, were above the 1901–1970 average values, and precipitation was reconstructed to be above average throughout the Northwest and in eastern North America and below average for southwestern North America. This pattern implies a strengthened eastward flow of air across the northwest United States and southwest Canada during the Little Ice Age with more frontal activity and precipitation along the United States–Canadian border. The higher pressures in the North American Arctic imply a predominance of Arctic air that could move southward into the American Plains behind trailing cold fronts embedded in the lows that were crossing the continent during this time period. The result would be severe conditions, increased storminess, and more moisture in the eastern and central parts of the continent.

A trend was noted in the importance of different storm tracks through the seventeenth through nineteenth centuries which could have been associated with a shrinking of the circumpolar vortex. During the seventeenth century, storms frequently traveled through the Pacific Northwest, but at times they entered California and traveled northeast along a California-Dakota track. Later, especially in 1831–1840, storms entered southern California more frequently and followed a southerly track into the eastern United States. This pattern becomes more and more important in the twentieth-century reconstructed climate but is probably associated with a subtropical jet stream and the displacement of the temperate jet stream to the north.

Little Ice Age temperature variations in western regions were also examined for the four seasons, but none showed marked cooling. If there is a Little Ice Age anomaly suggested by these reconstructions, it is one of higher moisture in the Northwest and cool temperatures in eastern North America. If the long-term temperature trends in these reconstructions are correct, advances in glaciers during this time period may be related more to high precipitation and accumulation of ice volume than to low temperatures and decreased ablation during the melting season.

Considerable climatic variability was reconstructed over this time period. The 1600s began as moist and cool with rising temperatures and declining moisture during the middle of the century and cooling at the end in western North America. Temperatures fluctuated throughout the 1700s in western North America but were warmer in the first half of the century than in the second half. The 1710s–1720s were the wettest, with generally drier conditions in the remaining part of the century. Temperatures peaked at the turn of the century and declined for the first 40 years of the nineteenth century, associated with increased volcanic activity and high moisture with maximum precipitation reconstructed for large areas of North America in the 1830s. Temperatures rose until the late 1860s and then declined to low levels early in the twentieth century in the western portions of the country. Warming began in other areas in the late 1800s. Moisture was generally low over the western parts of the grid during the last 60 years of the nineteenth century and higher during the first two decades of the twentieth century. Warming was evident everywhere by the 1930s.

The high sea-level pressure anomalies reconstructed over the eastern North Pacific and western North America during the Little Ice Age may have been associated with a weaker Aleutian Low or a strengthened North Pacific High with greater blocking of storms entering the California coast and drier conditions throughout southwestern regions. More storms appeared to enter Washington and British Columbia, however, enhancing precipitation there.

The reconstructions at the western margin of the climatic grid may have larger errors than those in the interior, because few good tree-ring chronologies were available from that region, and features in the circulation unique to the Pacific Coast may not have been captured by the reconstructions (see App. 2.16). The discrepancy between glacial advances during the Little Ice Age and the warm temperatures reconstructed for California, western Oregon, and western Washington may simply arise from errors of this nature. In addition, reconstructions in the area of the Aleutian Low did not verify as well as reconstructions in other areas (Fig. 6.3), leading to greater uncertainties about the exact mean position of the Aleutian Low between 1602 and 1900 (see App. 2.16).

9

Summary and Conclusions

A. Tree-Ring Data Sets

1. Grids of 89, 65, and 40 low-altitude, arid-site tree-ring chronologies (Fritts and Shatz 1975) from western North America were evaluated as potential sources of paleoclimatic information. The 65-chronology grid appeared to be the best compromise between the shorter and denser 89-chronology grid and the longer and less dense 40-chronology grid. Also, the climatic reconstructions from the 65 chronologies appeared to be superior.

2. Eight different species were studied and ten to twenty-five trees, two cores per tree, were sampled for each chronology. All annual rings were dated, and the nonstationary ring-width measurements were transformed into stationary time series by dendrochronological standardization techniques (Fritts 1976; Graybill 1979b). This transformation preserved most of the climatic variance in the tree-ring series at time scales shorter than 200 years but could have removed some of the variance for longer time scales. Otherwise, the climatic variance in the standardized chronologies appeared to be intact.

3. The shortfall of explained climatic variance in tree-ring reconstructions can be attributed to (1) nonclimatic factors present in the common tree-ring variance, (2) insufficient replication of sampling so that there are large errors in the chronology estimates, (3) too great a distance between the climate data and the tree sites, (4) omission of important climatic variables from the statistical analysis, (5) inadequate model structure, including nonlinear effects, (6) errors in the climatic data, (7) differences between the microsites of the climatic stations and the tree sites, (8) integration of the climatic record by the tree's responses, which smooth out and virtually eliminate a large amount of high-frequency variance, and (9) climate factors important to the tree that are not measured adequately by existing meteorological instruments (Pittock 1982, Fritts 1976).

4. The 65-chronology grid spanned an area of 4,620,000 km², including thirteen western U.S. states, three Mexican states, and two Canadian provinces. The density was 7.1 chronologies/Mkm², which was two to three times the density of other tree-ring data sets used to reconstruct spatial patterns of climate. All data in the 65-chronology grid spanned the seventeenth through nineteenth centuries and included at least the 1963 average ring-width index. The average chronology error at the beginning of the record was estimated to have been 250 percent of the error for 1901–1963, because only a small number of trees spanned the early part of the record. Estimates of climate at the beginning of the seventeenth century would be expected to have errors that are two and one-half times the errors in the twentieth-century portion of the record. By 1640 the errors are 200 percent, and by 1690 they are 150 percent. After 1800 the error is only slightly larger than the error of the twentieth century.

5. An approximate linear relationship between separation distance and the correlation between chronologies was evident. Fifty-eight percent of the variance was common to chronologies on the same site (after the effects of distance were assessed). The correlations were positive and significant up to an average distance of 1,154 km, and they were inverse and sometimes significant at a separation distance of about 2,000 km. This result suggests that both direct and inverse relationships with climatic variations outside the area of the tree-ring grid should be examined carefully and diagnosed.

6. The relationship with precipitation on these sites is primarily direct, and the relationship with temperature is primarily, but not exclusively, inverse. Although errors in these chronologies increase as a function of decreasing sample size in the early part of the tree-ring record, the errors are also a direct function of the size of the yearly index. Since low growth has less error than high growth, the estimates of either low precipitation or high temperatures would have less error than estimates of high precipitation and low temperatures. More estimates of drought or extreme warmth would be expected to pass significance tests than estimates of wet conditions or a cool climate.

7. The spatial patterns of variation are larger for sea-level pressure and temperature than for precipitation and tree growth, so fewer PCs are needed to describe the important spatial patterns in climate.

8. Since some of the spatial differences in the ring-width response to climate may be related to differences in species, exposure, altitude, and geographic location of the sites, a flexible multivariate model was selected that would be capable of transferring both the similarities and differences in tree-ring chronologies into spatial differences in precipitation, temperature, and sea-level pressure.

9. The chronologies were corrected for first-order autocorrelation by subtracting the product of the first-order autocorrelation and the departure from the mean for prior growth. Both the corrected and uncorrected chronology data were used as statistical predictors of climate in the calibration work.

B. Climatic Data

1. Ninety-six sea-level pressure grid points were selected at intervals of 10° and 20° covering latitudes 20° to 70°N and 80°W to 100°E (Blasing 1975). By the time the superior sea-level pressure set of Trenberth and Paolino (1980) became available, it was too late to repeat the experiment using this new pressure data set. The discrepancies between the Blasing and Trenberth and Paolino data sets were examined carefully, however, and the greatest differences were noted over Asia, the Arctic, and the Mexican highlands. It was concluded that only the data points along the northern, western, and southern borders of the 96-pressure grid were greatly affected by the Trenberth and Paolino corrections and that these parts of the grid would have to be interpreted carefully.

2. The variance in monthly average sea-level pressure over North America and the North Pacific was examined to determine the times of greatest variance change between seasons and to ascertain which months could be combined to make up each season. Winter was chosen as December–February, spring as March–June, summer as July–August, and autumn as September–November. The year began with December of the previous year and ended with November.

3. Grids of two sizes were selected for temperature and precipitation. The larger grids were used most extensively in this book. They extended well beyond the tree-ring predictor grid, which allowed for evaluating the maximum range of meaningful climatic relationships. The smaller grids were proximate to the tree-ring chronology grid and provided a measure of the maximum relationships to expect but omitted some important relationships, especially for temperature, with climatic variables beyond the boundaries of the tree-ring grid.

4. The temperature and precipitation records before 1901 were used for verification tests. These data were less reliable than the 1901–1963 data set used for the calibration work, and many observations were missing. Since all verifications for temperature and precipitation were based on these less reliable data, the verification statistics probably underestimate the real verification that exists.

C. Principal Component Analysis

1. Principal component and canonical regression analyses were used to identify and to maximize the relationships between large-scale differences in growth and climate. The smaller PCs and insignificant canonical variates were eliminated from the analysis to minimize the smaller-scale differences that were likely to be noise, local nonclimatic disturbances, microclimatic variations, or simply error in the measurements. The most important PCs of spatial temperature, precipitation, and sea-level pressure variation for each season were calibrated with the most important PCs of spatial chronology variance.

2. Eigenvectors and their PCs were extracted from the 65 tree-ring chronologies (1601–1963) with first-order autocorrelation (matrix I) and with first-order autocorrelation removed (matrix M). The largest 15 were selected and used as statistical predictors of climate. The PC values from the preceding year (B), the immediate year (I), the following year (F), and 2 years following growth (FF) were used so that lagging, autoregressive, and moving average relationships between climate and growth could be assessed. The first 15 PCs of B, I, F, and FF represented 69 percent of the chronology variance. The first 15 PCs of the M matrix represented 67 percent of the corrected chronology variance.

3. Spectra of the variance summed over the largest PCs of the I series exhibited the typical red noise characteristic of many climatic data sets. The spectra for the M PCs had somewhat more variance at time scales of 2 to 3 years than at time scales of 3 to 20 years. When the data were fully ARMA modeled, no differences between these time scales were evident. Only small differences in the reconstructed maps could be identified between the M and the fully modeled chronology sets, so the original strategy was adhered to in the remaining work to correct for only the first-order autocorrelation effects. Future work should use ARMA-modeled residuals at the very beginning instead of the M data set.

4. The 2 most important PCs of tree growth had 10 to 11 percent more variance in portions of the record from the twentieth century than in those from the seventeenth through nineteenth centuries. Such differences in growth could lead to comparable difference in reconstructed climatic variance, depending on the coefficient assignments in the canonical regression analysis.

5. The 15 largest PCs of sea-level pressure (1899–1970) and temperature (1901–1970) and the 20 largest PCs of precipitation (1901–1970) were considered in this analysis. There was no simple answer as to how many climate PCs were important enough to be calibrated. It seemed

best to start with a fairly high number and to eliminate the unimportant relationships by rejecting and excluding the canonical variates that contributed insignificant variance to the regression of climate on tree growth. Model structures with 2 to 20 candidate PCs of climate were calibrated and evaluated.

D. Modeling the Tree-Growth Response

1. Response-function results were used to ascertain the sign and strength of the linear correlation of monthly temperature and precipitation with the chronology growth index. This procedure allowed for screening a large number of possible climatic variables, helped to choose which monthly variables should be combined into seasonal values, and indicated the relative importance of temperature and precipitation to tree-ring growth on the different sites.

2. The response functions indicated that local conditions of precipitation appeared more tightly coupled to variations in ring width than to local conditions of temperature. Also, the climate during the prior summer, autumn, winter, and spring was as important to ring-width variations as the climate during late spring and summer concurrent with the season of growth. Both the sign and size of the coefficients varied from season to season, but many similarities in response-function weights were evident and appeared to be associated with similarities in species, exposure, altitude, and geographic location of the sites.

3. Some of the response-function differences undoubtedly resulted from errors, random variations, and unmodeled relationships. Others may be due to variations in response to the same climatic variables from one season to the next. It was hypothesized that this latter information might be used if a flexible calibration procedure could be employed to accommodate both the differences and similarities in these chronology responses.

4. Possible modeling structures that were considered to handle the biological and statistical features of these data included (1) lagging ring width one or more years behind the climate, (2) incorporating the climate of the prior year in the modeled relationship, (3) removing first-order autocorrelation from each chronology data set, (4) relating growth to the climate in each season rather than to annually averaged climate, (5) using principal components to emphasize the large-scale features of the relationships, and (6) employing canonical regression to provide flexibility in the assignment of the transfer-function weights.

5. Bates and Granger (1969) show that the reliability of a statistical forecast may be improved by averaging statistical estimates from two or

more calibrations using different combinations of predictor variables. Therefore, models of varying structure were calibrated, verified, and the results examined for improvement. The combinations of two to three models with significant improvement and with the best statistics were selected. The best combinations for each of the four seasons were averaged to obtain annual estimates.

6. The relationships considered in this study have been averaged over two cores per tree and over many trees and many years. It was assumed that by the time these relationships have been integrated over seasons and generalized as PCs of both tree-ring series and climatic data, most curvilinear relationships were reduced to more or less linear form, and non-normal characteristics of individual data points became less important. Multivariate calibration models were applied to only 61 to 63 years of climatic data (1901–1963). At this stage in the analysis it seemed more reasonable to model lags, autocorrelation, and differences in response over space than to consume *df* by modeling curvilinear effects.

E. Transfer-Function Model Development

1. Canonical regression was used to calibrate the PCs of a climatic variable for one season with the PCs of the 65 tree-ring chronologies. The calibration and verification statistics from a variety of models with different structures were used to decide which model structures were most appropriate. Fifteen or 30 predictor PCs of the tree-ring chronologies were found to be optimum, and these numbers were then used as predictors of 2 to 20 PCs of climate. The structure of the equation was varied by including different lags and numbers of candidate PCs of climate in the canonical regression analysis.

2. An *F* statistic, corrected for the autocorrelation of the residuals, was used to test and retain only those canonical variates and their associated PCs that contributed significantly to the regression variance. Generally, the larger the number of candidate climate PCs available for analysis, the larger the number of significant canonical variates that could be examined and entered into regression.

3. The actual number of canonical pairs selected for the transfer function varied from one analysis to the next. They were not necessarily the pairs with the highest canonical correlations but rather the pairs that reduced the most climatic variance.

4. All available temperature and precipitation data before 1901 were used in the verification. The independent data for the seasonal sea-level pressure were generated by subsample calibration. Five verification

statistics were calculated, and their significance was tested. In addition, the reduction of error and its three components were calculated, and the coefficient of contingency, skewness, and normality of the dependent and independent data were estimated.

5. More than five hundred calibrations and verifications were calculated, and all models with more verification tests significant than would be expected by chance were ranked using seven calibration statistics and seventeen verification statistics. It was assumed that those models with the highest ranks produced the most reliable reconstructions, but when there were ties, the simpler models or the models with the most reasonable structure were selected.

6. A number of the highest-ranking models with differences in structure were selected from each variable, season, and grid. Combinations of two or three high-ranking models were merged (averaged) and their statistics calculated; these data were ranked, and the models with highest-ranking statistics were chosen. The reconstructions from the best seasonal estimates for each climatic variable were combined to obtain the annual estimates.

7. Merging, on the average, appeared to result in improvement. The low-frequency variations, particularly those for sea-level pressure, appeared to be more reliable than the high-frequency variations, although limited df of the low-frequency variations prohibited rigorous evaluation of their statistical significance.

8. The df of the model residuals ranged from 28 to 57 out of a total of 61 to 63 years used in the analysis. The df for most models ranged from 40 to 50. This number appeared to be adequate for the verification testing, except for the filtered low-frequency features at time scales of 10 years or longer.

9. The collective significance of the grid-point data was also evaluated, using the guidelines described by Livezey and Chen (1983). The critical threshold was low for all of the reconstructions and well below the statistics of the final reconstructions. This result indicated that the grid-point statistics were well above the expectations for chance variations.

F. Characteristics of the Reconstructions

1. The 46-grid temperature reconstructions exhibited the highest explained variance, and the 96-grid precipitation reconstructions exhibited the lowest explained variance.

2. Areas of high calibration and verification of sea-level pressure were the Canadian Arctic, the southwestern United States, and large

portions of the North Pacific. The poorest calibration and verification were for summer, when smaller-scale patterns were more evident, especially in the area of the Aleutian Low and along latitude 50°N from the western to the eastern border of the grid. This poor calibration was in the area of large pressure variations associated with the succession of cyclones and anticyclones that are carried in the upper-level flow near the zone of confluence (Lamb 1972).

3. The reconstructions of temperature passed verification tests and appeared reliable for some areas in the Plains states, western Great Lakes, and upper Mississippi Valley; verification was poor in some areas of the Southwest and central western states. The annual temperature data verified the best, but three areas of poor reconstruction were indicated in the Intermontane Basin of the West, the Ohio River Valley to the Atlantic Coast, and an area stretching from southeastern Arizona through the Gulf states to the Atlantic Coast. The shortness of the independent record from many western stations (Figs. 3.2 and 3.3) helped to explain some of these differences. Calibration and verification was better than would be expected by chance over a number of distant stations, some of which are 1,000 to 2,000 or more km distant from the tree-ring chronology grid. Therefore, it was decided to evaluate the results for all stations in the larger grids.

4. Good precipitation reconstructions for seasons were largely confined to the western states and to winter and spring. Although the area of reliable annual precipitation reconstructions was more dependent on the proximity of the tree-ring grid than on annual temperature and sea-level pressure, the statistics suggested that some reliable information on precipitation may have been reconstructed to the east of the tree-ring grid.

5. The spatial variations in temperature and sea-level pressure were reconstructed more accurately than the spatial variations in precipitation. This strong linkage between arid-site tree growth and spatial variations in temperature and sea-level pressure may be attributable in part to the higher spatial correlation for temperature and sea-level pressure than for precipitation. For example, a temperature record was correlated with more tree-ring chronologies than a precipitation record simply because temperature has more large-scale variance to be calibrated. This result is supported by the observation that the first few eigenvectors of temperature, sea-level pressure, and tree growth reduce more variance than the first few eigenvectors of precipitation and that, usually, fewer canonical variates were significant in the regressions of precipitation than in the regression of the other two variables.

6. The calibrated variance percentage generally increased (1) when the reconstructions from two to three optimal seasonal models were

averaged, (2) when the grid-point data from the averaged models were pooled to form regionally averaged climatic estimates, (3) when the regional data were filtered before comparison, and (4) when the seasonal data were combined to obtain annual values. The greatest increases in calibrated variance occurred when the seasonal merged models were combined into annual averages and when the annual averages were summarized as regional averages and filtered. This result supports the finding that potentially more information may be present in the large-scale patterns, and it suggests that low-frequency portions of the regionally averaged annual climatic reconstructions might be more reliable than the higher frequencies. Nevertheless, wide variability in the quality of these low-frequency reconstructions was noted.

7. Spectral analysis showed that the low-frequency variations, particularly those for time scales of 20 years or longer, were well reconstructed and verified for sea-level pressure and temperature. The 5 largest PCs of annual sea-level pressure and temperature were well calibrated, whereas only 3 or 4 of the largest PCs of annual precipitation were well calibrated.

8. After all adjustments were made, 42.9 percent of the unfiltered variance of annual regional climate was calibrated over all regions, and 58.5 percent of the filtered variance of annual regional climate was calibrated. The filtered regional calibrated variance averaged 62.1 percent for annual temperature and 70.6 percent for annual sea-level pressure. The calibration statistics were higher for some regions than for others, ranging up to 80 percent for filtered temperature estimates. These results compared favorably with those from other dendroclimatic studies that have been reported.

9. The analysis of calibrated fractional variance for seasons indicates that, in general, the spring reconstructions were superior. The individual components for winter were as well calibrated as those for spring, but the calibration statistics for the regional averages and low-frequency variations for winter were not as high as those for the spring season. The calibrated variance for low-frequency variations over regions for summer was higher than that for the low-frequency variations for any other season, but the statistics for the unfiltered reconstructions were lower than those for winter or spring. The calibration for autumn was clearly the lowest of the four seasons.

10. The considerable variability noted in the spatial patterns of both calibration and verification statistics for the different grids, climatic variables, seasons, and regions was expected for the following reasons: (1) large differences were noted in the response of the trees to the climate in different seasons, (2) the distribution of both the tree-ring and the climatic data points was uneven, and (3) the independent

temperature and precipitation data sets were of poor quality, with a very uneven distribution and variable length, especially in the western parts of the temperature and precipitation grids.

11. Furthermore, the high variability between regions can be attributed to other possible sources, including (1) natural random variability, (2) variability in the forest-stand responses, (3) differences in trends between the climate in the mountains supporting the forest stand and the climate in the valleys where the climate stations are located, (4) seasonal variations in the climatic features of the regions, and (5) interactions between the response of the trees and variations in climate in the different seasons.

12. Pooling of data over the grid points within climatic regions did appear to improve calibration consistently. Regions 2, 4, 6, and 7 appeared to have the most reliable low-frequency estimates of temperature.

13. Generally, the verification statistics supported the calibration results. A greater number of verification tests passed with increasing levels of pooling. The annual values were more reliable than the seasonal values. The reconstructions for different seasons had variable characteristics, and those for winter, spring, and summer were the most reliable. The parametric statistics were more often significant than the nonparametric statistics.

14. The adjusted fractional variance calibrated in the smaller temperature and precipitation grids was, on the average, 0.06 to 0.07 higher than that for the larger grids. When only the six western regions of the 77- and 96-grids were considered, the differences became less pronounced and sometimes reversed. Regions 7–9, however, which were east of the chronology grid and were available only from the large grids, had significant statistics with values that were sometimes as high as or higher than those in the smaller grids. In addition, sea-level pressure reconstructions for areas well to the north and west of the tree-ring grid often had significant statistics. Although the average performance of the smaller grids always surpassed that of the large grid, some important climatic linkages with regions to the east, west, and north would not have been identified if only the small temperature and precipitation grids had been included in the design of this analysis.

15. It was concluded that the calibration results for regions 1–6 in the West were comparable for both temperature grids. The temperature reconstructions for regions 7–9 are also good, but no reconstruction of any variable was found to be reliable in regions 10 and 11 in the East. The highest-quality reconstructions of precipitation were those from the small grid. Those precipitation reconstructions from regions 1–6 in the large grid were acceptable but of lower quality. The only accurate

precipitation reconstructions outside the tree-ring grid were those for a small number of stations in regions 7–9 to the east of the tree sites. The quality of these reconstructions, however, was inferior to the quality of precipitation reconstructions for many stations in the West.

16. The reconstructions in the western parts of the large grid were so similar to the reconstructions from the small grids that it was decided, for the diagnostic purposes of this book, to focus on the reconstructions from both the eastern and western portions of the large grids. This process would allow evaluation of stations both upwind and downwind from the tree-ring chronology grid. The reliability of all data beyond the boundaries of the tree-ring grid was critically examined, however, and the greatest uncertainties in the reconstructions are acknowledged in the text.

17. The discussion of the reconstructions focused on the large-scale, annual, and low-frequency climatic patterns, because they appeared to be reconstructed most accurately.

G. Large-Scale Reconstructions through Time

1. The reconstructions were used to relate the twentieth-century instrumental measurements to the estimates from the seventeenth through nineteenth centuries. More variance existed in the annual reconstructions for the seventeenth through nineteenth centuries than in those for the twentieth-century record. However, large mean changes in reconstructed climate in the twentieth century appeared as an increase in low-frequency variance. Annual temperature estimates also showed increases in variance in the twentieth century for time scales of 8 years or less. The low-frequency variations in temperature were most important in regions 5–9 east of the Rocky Mountain crest.

2. The greatest reconstructed mean temperature change was the warming that occurred in the twentieth century. This rise in temperature became apparent in the eastern United States late in the nineteenth century, with the greatest 30-year mean temperature change occurring in 1918 and the most rapid increase in the variance of 30-year overlapping means beginning in 1911. The highest 10-year mean temperature for the 77-station grid began with 1930 and was 0.55° C higher than the 1901–1970 mean value. Temperatures were reconstructed to have been exceptionally low during the first two decades of the twentieth century in the western portions of the grid, though, and precipitation was high in the West and moderate in the eastern parts of the grid. Severe drought was evident in the 1930s, and a less severe drought was apparent in the 1950s and 1960s.

3. For the first three decades of the seventeenth century the average reconstructed temperatures in the West were near their mean value, and then temperatures rose rapidly, reaching highest values in 1657–1666. By 1667 temperatures were in rapid decline in the West, down to levels below those of the twentieth century. Warming was reconstructed in the 1680s, and cooling was reconstructed in the 1690s. These temperature reconstructions were more similar to the model results of Robock (1979), which used volcanic dust for forcing, than to the results of Groveman and Landsberg (1979), which Robock used for comparison, although some divergent trends were apparent.

4. The 1610s and 1630s were reconstructed to have been exceptionally wet in the West, but conditions were drier and sometimes well below average for the remainder of the century. The decades beginning in 1663 and 1682 were among the five driest noted.

5. The period from 1650 to 1720, the Maunder minimum, when sunspot numbers were especially low (Eddy 1976, 1977), does not stand out as a period of anomalous climate. A cooling trend was evident in the West, and a warming trend in the East was interrupted by cooling in the 1670s, 1695–1704, and the 1710s. These periods were also punctuated by four to six droughts. The oscillations in temperature and droughts were consistent features in both the large- and small-grid reconstructions, and no intervals with high precipitation were evident. Similar features can be seen during other time periods, but they were lower-frequency events.

6. Temperature reconstructions during the eighteenth century fluctuated about the twentieth-century means. The warmest decade in the West was 1717–1726; the coolest was 1760–1769. The amplitude of the temperature variations was low in the third quarter of the century. Temperatures began to rise and the variability increased in the 1780s, reaching maximum values around the turn of the century. Except for the high temperatures at the end of the century, similarities with the Robock (1979) model using volcanic dust forcing and with Lamb's dust-veil index (Lamb 1970) suggest that prolonged periods of high temperatures were associated with time periods of low volcanic activity and low dust-veil index. Little agreement was found between the data of Catchpole and Faurer (1983) and Guiot (1987) for eastern Canada, probably because of the great distance between that area and the tree-ring grid.

7. Precipitation was reconstructed to have been low at the beginning of the eighteenth century. Maximum values were reconstructed for 1718–1727 at a time when Cook et al. (in press) reconstructed high moisture in the southeastern United States. The remaining decades were relatively dry, with the most severe droughts in the 1750s and

1770s, and droughts were reconstructed in the East during the latter period (1769–1776) by Cook et al. (in press).

8. There were only two peaks and troughs in the temperature variations of the nineteenth century. High temperatures at the beginning of the century declined, reaching low levels in the 1830s, but the lowest temperature estimates were for the less reliable stations in the East. The widespread cooling was associated with increasing volcanic activity, large dust ejections, and cooling of the earth, although immediate cooling by Tambora in 1816 and 1817 was not reconstructed in the West. Warming began in the 1840s, with maximum warmth reconstructed for the 1860s. Temperatures declined in the 1870s and 1880s and were associated with volcanic activity and dust-veil events. Temperatures in the East were reconstructed to remain low until the twentieth century, but they were not as extreme in western regions.

9. Precipitation declined over the first decade of the nineteenth century and then rose to all-time highs in the 1830s for many western stations. Similar features were reported in the eastern United States by Cook et al. (in press), with drought in 1814–1822 and very wet conditions in 1827–1837. This decade appeared to be the wettest interval from this 360-year record accompanied by cooling climate. This wet period was followed by low reconstructed precipitation in the 1840s and 1860s. There was less variability in precipitation for the remainder of the century, with a gradual rise in precipitation toward the end.

10. East-west seesaw variations in temperature and north-south variations in precipitation were noted, which were similar to the eigenvector patterns obtained from the instrumental record. Precipitation in the Northwest was often out of phase with the precipitation variations to the south. Greater differences between the two precipitation grids than between the two temperature grids reflected the lower reliability of the precipitation results.

11. In the seventeenth century, precipitation was reconstructed to have been high in the Columbia Basin and lower in the other five western regions. In the eighteenth and nineteenth centuries the difference was less but continued and higher precipitation in the Northwest was identified as one likely characteristic of the Little Ice Age climate.

12. Many episodes of low precipitation were reconstructed for the ten grid points in the central Plains of North America, which were the only ones for that variable with reliability beyond the boundary of the tree-ring grid. Many intervals were reconstructed to have been cool, however, when the adverse effects of low precipitation may have been ameliorated to some extent. Three out of four droughts noted by Stockton and Meko (1975) were also periods reconstructed to have low precipitation and high temperature. Stockton and Meko's conclusions

are supported in that drought conditions reconstructed between 1700 and 1900 appear not to have been as severe as the drought of the 1930s, but some of these droughts were longer-lasting events.

13. The temperature and precipitation reconstructions suggest that, if the climate continues to warm as predicted (Hansen et al. 1988) and precipitation is at least as variable as it was during the last three centuries, the projected rising temperatures would be likely to intensify the effects of low precipitation, making future droughts in the central Plains more sustained and severe than observed or estimated.

H. Spatial Patterns of Reconstructed Climate

1. The features of an embedded pattern in the reconstructions for 1602–1900 (Fig. 7.6) as compared to averages for 1901–1970 may reflect characteristics of the Little Ice Age climate. They include (a) sea-level pressure above the twentieth-century values, with the greatest anomaly over eastern Alaska and northwestern Canada (implying that the Aleutian Low was displaced west of its twentieth-century position or that the North Pacific High was northeasterly displaced); (b) annual temperatures higher than the twentieth-century average for the entire western mountainous region but lower than the average east of the Rocky Mountains to the Atlantic Coast; (c) annual precipitation higher than the twentieth-century average from the Pacific Northwest to the Great Lakes and for the entire eastern half of the United States, and lower from California to western Wisconsin and southward to Kansas, Oklahoma, and east Texas.

2. A synoptic interpretation of these differences suggests that in the last few centuries of the Little Ice Age a greater number of storms may have been deflected north of their twentieth-century mean position through the central North Pacific into Washington and British Columbia. This pattern would have produced an anomalous airflow from the south with warmer temperatures in the West and higher precipitation in the North as more storms with associated fronts traveled through the United States–Canadian border states in the West. Anomalously high sea-level pressure in northwestern North America and lower anomalies to the east also suggest that the Icelandic Low was strong or westerly displaced with more cold air outbreaks east of the Rocky Mountains and cooler temperatures in the East. Fewer storms moving along the southerly track through the Southwest than in the twentieth century would be consistent with the low precipitation reconstructed in that part of the grid. Enhanced precipitation in the Pacific Northwest

suggests that higher precipitation rather than lower temperatures may be responsible for glacier expansion in the Little Ice Age for northwestern areas (Denton and Porter 1970; Grove 1988). An expanded circumpolar vortex is implied by the higher sea-level pressures reconstructed in the North American Arctic.

3. This anomaly, which is embedded in all reconstructions for the seventeenth through nineteenth centuries, should not be considered an unwanted bias or problem with the reconstructions. It simply reflects the climate of the seventeenth through nineteenth centuries compared to the large anomaly in the twentieth-century climate.

4. The seventeenth century began with a near-normal climate followed by a low-pressure trough over western North America in 1611–1620 with high precipitation and northward flow of a warm but unverified anomaly of tropical air into the East. High-pressure blocking patterns in 1621–1630 brought drier conditions to the Southwest as more storms may have traveled along northern tracks with the northerly flow of warm air into large areas of the United States.

5. The first significant mean temperature change was reconstructed around 1637 and lasted until 1666. Temperatures increased in the West and declined in the East as a strong Aleutian Low in the central North Pacific and an Icelandic Low may have favored the northward flow of warm air into the West and southward flow in the East. Precipitation varied in the West with the strength of the Aleutian Low, and temperature was inversely related to the strength of the Icelandic Low.

6. Anomalously high pressures dominated the North American Arctic and the North Pacific from 1667 to 1716, when the Maunder minimum was evident. Associated features were a weakened Aleutian Low or strengthened North Pacific High, declining temperatures throughout the United States, declining moisture especially in the central Plains and East, and a corridor (referred to as the *California-Dakota storm track*) of increased precipitation in the West. Storms may have been deflected to southerly tracks across the eastern North Pacific, reaching land more often along the California coast, where they traveled in a northeasterly direction north of the Great Lakes. Though the temperature change was downward, temperatures relative to the twentieth century were higher in the West than in the East.

7. Sea-level pressures declined in 1717–1760 for large areas of the North Pacific, with weak anomalies from the twentieth-century values evident. Temperatures were less extreme, especially east of the Rocky Mountain crest, and moisture declined in the California-Dakota track but was higher elsewhere on the map. During the 1730s through 1750s warmth in the West was followed by drier conditions throughout large parts of the United States.

8. Higher pressures returned to the North Pacific and Canadian Arctic in 1761–1790s; more storms may have followed the California-Dakota storm track, but the changes were not as marked as in 1667–1716. Temperature anomalies were reconstructed to have been lower in the East than in the West. Dry conditions were reconstructed primarily in the Southwest and South, although Cook et al. (in press) report that dry conditions were reconstructed in 1769–1776 for the East.

9. The 1791–1820 period was reconstructed to have been the warmest interval throughout the United States, exceeded only by the warming of the twentieth century. Volcanic activity was average. Pressures that declined over the North Pacific and the western North American Arctic suggest the presence of a stronger and northerly displaced Aleutian Low with more storms moving into the Arctic. Temperatures rose throughout the United States, drought was pervasive in the Great Plains states, blocking patterns were more common, and storm movement through the California-Dakota storm track may have been reduced. Severe drought was indicated and verified in the East. Temperatures declined but, even with the great eruption of Tambora, were not substantially lower than the twentieth-century means in the West.

10. The most rapid and marked decline in 30-year mean temperature for the reconstructed grid was noted for 1821–1849. It was associated with increasing volcanic activity, rising sea-level pressure in the eastern North Pacific, and declining pressure over Canada that suggested an enhancement of trough development in the eastern Canadian Arctic. The greatest cooling was reconstructed in Saskatchewan near the point of maximum trough development. Decreasing amounts of precipitation were reconstructed in the Northwest, and increasing amounts of precipitation were reconstructed in the Southwest.

11. The 1831–1840 decade was extremely moist for large areas of the United States, and this condition was confirmed in the East (Cook et al. in press). High-pressure anomalies reconstructed over the Klondike may be associated with cold temperatures reconstructed east of the Rocky Mountains. More storms may have traveled inland south of the former path and south of the California-Dakota storm track, through the southwestern Plains and into the Northeast. This pattern would advect large amounts of moisture from the Gulf of Mexico that would produce high precipitation as warm gulf air met cold air from the north.

12. Temperatures were higher in 1850–1876, sea-level pressure declined slightly in the eastern North Pacific and rose in the eastern Canadian Arctic, and drought was reconstructed for many parts of the United States.

13. Sea-level pressures increased over Alaska and the western Canadian Arctic around 1877 and decreased along the Pacific Coast of the United States. This condition more or less persisted until 1917. More cold air was advected into the central United States, and temperatures decreased everywhere but in the Southwest. More storms may have traveled inland over central California, and increased precipitation in the West suggests that storms may have traveled more frequently along the California-Dakota storm track. Eastern temperatures began to increase in the 1890s but declined in the West, perhaps associated with high sea-level pressure anomalies in the North American Arctic and North Pacific. Lower sea-level pressure off the Pacific Northwest early in the twentieth century was associated with more moisture and cooler conditions in the West.

14. After 1920, warming was evident throughout the United States. Sea-level pressures declined as more storms traveled into the Arctic. With the shrinking of the circumpolar vortex, the subtropical jet may have increased in importance and enhanced the flow of warm moist air into the United States. The patterns in the reconstructed anomalies were weak but similar to patterns that were found in the instrumental measurements for the six decades when both records were present.

I. General Conclusions

1. Consistent and reasonable patterns among the three variables plotted on the maps support the conclusion that the reconstructed spatial anomalies in climate portray relationships that actually exist. They also suggest probable airflow patterns, storm tracks, and high-pressure blocks.

2. Often annual temperature variations in the western United States appear to be out of phase or unrelated to worldwide changes. The annual temperature reconstructed for 1602–1900 was cooler in the East and warmer in the West than the 1901–1970 mean climate. Seesaw variations between the eastern and western United States are common. The early portion of the twentieth century was reconstructed to have been cooler than the seventeenth through nineteenth centuries for five regions in the West and warmer for other regions in the United States.

3. The reconstructed record suggests that the twentieth century was considerably wetter in the Southwest and drier in the Northwest than it was during the seventeenth through nineteenth centuries. Thus, the long-term projection for hydrologic conditions in the Southwest based on these reconstructions would indicate drier conditions than the shorter instrumental record indicates.

4. The broad-scale patterns of reconstructed temperatures were compared to other well-known climatic series to see whether the reconstructed temperatures were consistent with the regional temperature variations observed by other workers. Direct correlations were noted with the hemispheric-wide temperature changes of Jones et al. (1982) from 1881 to 1946, but this relationship weakened beginning in 1947. Significant inverse correlations were noted from 1748 to 1756 with Manley's central England temperatures (Manley 1974), and no other period was significantly correlated. The temperature reconstructions from North America were not related to Manley's temperature record.

5. The temperature averages were significantly correlated with the Groveman and Landsberg hemispheric temperature estimates (Groveman and Landsberg 1979), but there were disagreements, largely during times when reliable proxy information from the North American area was absent from the Groveman and Landsberg data set. It was suggested that the Groveman and Landsberg estimates might be improved substantially by using the continuous series of average temperatures reconstructed for the 77-grid in place of the more discontinuous and less reliable North American proxy data used in their work.

6. Significant and consistent relationships were noted between the Southern Oscillation (SO) and the patterns of temperature and precipitation from both the instrumental record and the reconstructions. This consistency with the SO data confirms the integrity of the reconstructions of precipitation and temperature. The relationships with tree-ring data involve only the temperate-climate teleconnections; no tree-ring data were available from the tropical regions, so the stronger tropical linkages were left out (Lough and Fritts 1990).

7. The long reconstructed record of climatic variation was related to dust veils from past explosive volcanic eruptions at lags of up to 3 years (Lough and Fritts 1987). Significant cooling was noted after major low-latitude eruptions for large areas of the United States, particularly in the central Plains. Nonetheless, winters were significantly warmer in the western states and summers warmer in the Far West for 1 to 2 years after major low-latitude eruptions. Cooling appears not be a universal response to low-latitude volcanic forcing, as warming can occur, particularly in the North American West. Longer-term cooling effects are implicated, but the significance of these relationships could not be tested.

8. Temperature reconstructions derived from arid, lower forest-border chronologies are somewhat different from those derived from upper tree-line, high-altitude chronologies. These differences may be explained by (1) seasonal variations in the response to temperature, (2) seasonal variations in the response to precipitation and soil mois-

ture, and (3) varying effects of other factors such as winter snowpack, cloud cover, wind, and humidity. The differences suggest that high-altitude chronologies and alpine glacial evidence should be interpreted as a response to warm-season temperatures and perhaps to cool-season moisture, not to annual temperatures as reconstructed from the low-altitude, arid-site dendrochronologies.

9. The conclusions about climatic variations from 1602 to 1960 are offered as tentative hypotheses derived from one dendroclimatic analysis and test using 65 chronologies from western North America. These conclusions must be compared to data from other independent paleoclimatic sources that can reveal changes on seasonal and decadal time scales with yearly precision. The technology and approach described in this book should be applied to a larger and denser tree-ring data set to obtain more precise reconstructions of climatic variables and to extend the reconstructions to cover larger regions about the earth. The only requirements are that a grid of datable, reliable, and climatically sensitive tree-ring chronologies of suitable quality and record length exists and that a network of climatic stations with long and reliable records of past climatic conditions is available for the work of calibration and verification.

10. This book examines the kinds of information that can be extracted from a spatial grid of well-replicated and dendrochronologically dated ring widths from arid, lower forest-border sites. As in all studies, the conclusions are limited by the research strategy used in this diagnostic test (see App. 1, Sec. H). The study focused on climatological variables of annual temperature, precipitation, and sea-level pressure. The strategy of analysis did not allow changing the seasons and grids or combining variables to obtain a "better" environmental index of the growth and environmental response (Stockton and Meko 1983; Cook and Jacoby 1977, 1983). It did provide an evaluation of where the temperature, precipitation, and sea-level pressure estimates are best and what might be done to improve the estimates.

11. The most important conclusion is that the large-scale temperature reconstructions are more reliable than the large-scale precipitation reconstructions obtained from a grid of arid, lower forest-border sites (see Figs. 6.8, 6.10, 6.15; Table 6.5). This conclusion contradicts the popular notion that arid-site trees respond only to precipitation and drought (Stockton et al. 1985; Schweingruber 1988) but is consistent with numerous analyses indicating that both temperature and precipitation relationships are important on arid sites (Fritts 1974, 1976; Brubaker 1980). Although it is usually acknowledged that temperature affects growth through water stress (Stockton and Meko 1975, 1983), the importance of temperature on other physiological processes (Fritts

1976) is often underestimated or not addressed. The greater spatial coherency of temperature over that of precipitation is also an important contributor to this particular result. The large-scale tree-growth patterns resemble the large-scale temperature patterns more than the large-scale precipitation patterns. At the very least, it should be acknowledged that temperature can be as important as moisture to arid-site tree growth.

12. Temperature estimates from arid-site chronologies may be confounded by collinearities with the precipitation response. Therefore, it is necessary to compare the temperature estimates with reconstructions from other sources. The temperature record reconstructed from high-altitude and high-latitude dendrochronologies may also be confounded by the effects of other climatic conditions. Both kinds of reconstructions should be reevaluated carefully in terms of their similarities and their differences.

13. The reconstructions from this work have been mapped and saved on microfiche and are available in digital form on magnetic tape or floppy disk. Fritts and Shao (in press) have developed IBM-PC programs and data sets that allow tree-growth patterns and the spatial reconstructions of temperature and precipitation for any combination of years from 1602 to 1961 to be mapped. A user's manual for the mapping utility is included here as Appendix 3. The utility and reconstructed data are available from Bruce Bauer, Data Manager, National Geophysical Data Center, 325 Broadway (E/GC), Boulder, Colorado 80303, USA, and the University of Arizona Press, 1230 N. Park Avenue, Tucson, Arizona 85719.

10

Recommendations

1. Canonical regression is an efficient procedure for relating and transferring large spatial fields of well-dated tree-ring chronologies to large spatial fields of climate. The procedure is relevant to multivariate problems involving a number of predictand variables, and an F test can be used to eliminate the canonical variates that contribute insignificant variance to the predictand estimates.

2. Large-scale eigenvectors of these spatial fields and their principal components can be used to identify the most important modes of variation, reduce the number of variables, and enhance the amount of signal over noise in the data sets.

3. The tree-ring chronologies should be fully ARMA modeled before obtaining the eigenvector matrix. The PCs can be lagged in time to model lagging relationships not removed in the ARMA-modeling analysis. However, the predictand variables can also be ARMA modeled (Meko 1981; Box and Jenkins 1976), the PCs computed from the residual time series, these PCs calibrated with the PCs of climate, and the coefficients of the transfer function applied to the predictor PCs to estimate the PCs of climate. Then these estimated PCs can be multiplied by the eigenvectors, and the ARMA-model coefficients can be applied to convert them to estimates of the original climatic variance.

4. The F test on each canonical variate provides an objective selection procedure that eliminates information in the PCs that is not a part of the covariance matrix. This test appears to be superior to arbitrary selection of PCs based on the size of their element in the eigenvalue matrix. The F test is not identical to the test used by Cook et al. (1988) to simulate this work. This difference between the two methods should be tested thoroughly before comparability of the test results is claimed. The canonical regression appears to work best if enough candidate PCs (in this case 5 to 15) are entered in regression to allow a number of canonical variates to be calculated and tested. For this reason, more than the minimum number of PCs should be used as potential predictors and predictands, and the F statistic should be used to select and

reduce the choice to a small number of canonical variates expressing only the important modes of covariance. Such a strategy assumes that a large number of trivial PCs are not considered and that enough *df* are left for statistical testing after the significant canonical variates are selected.

5. It may have been better to use response-function results to determine the seasons to be calibrated than to use climatological considerations and statistics of climatic data to choose the seasons for this analysis. In addition, if three-month seasons had been used throughout the analysis, it would have been easier to compare the findings to those from studies using the conventional seasons.

6. The correction for the arbitrary use of subsample 1 in the sea-level pressure calibration (Chap. 5, Sec. G) is probably incorrect. The difference between the means of the independent data and the mean of subsample 1 may have provided a more realistic error estimate and a more realistic *BIAS* statistic.

7. Temperature and sea-level pressure can be reconstructed for a number of points outside the grid of tree-ring chronologies, but precipitation reconstruction is more limited, although some points to the east of the tree-ring grid appear to have reasonable estimates. Not all climatic grid points within the area covered by the tree-ring chronologies could be calibrated and verified. All reconstructions for points outside the area of the tree-ring grid need to be carefully analyzed and validated with independent climatic information.

8. It is always desirable to have climatic information from a variety of proxy records, and it is still necessary to search for and assemble more proxy data sets. Proxy evidence of large-scale, high-frequency climatic variations needs to be examined from a climatological viewpoint to evaluate both differences in the various climatic sensors and differences in regional and global climate. Few well-dated proxy sets other than tree-ring data have been assembled for the American West, however, and none has been evaluated for its climatic information content. Each proxy datum has its own error of estimate, and when two proxy data are compared, the two errors can have a compounding effect. Agreement between proxy data is an encouraging result, but disagreement does not necessarily indicate that a particular climatic estimate is incorrect. At present, tree-ring data are about the only reliable source of climatic information with accurate dating, sufficient resolution, and adequate spatial coverage to reconstruct annual variations in climate over a wide spatial field for a region like North America.

9. The ring-width responses to climatic variables can differ markedly from one season to the next. This variation can be dealt with by calibrating and verifying variables for different seasons and combining the seasonal reconstructions to obtain the annual estimates.

10. Averaging may increase the signal-to-noise ratio of a reconstruction by reducing the randomly related differences and errors of measurement. Reconstructions from two or three models of different structure can be averaged, the grid-point estimates within homogeneous climatic regions can be averaged, and these data can be filtered or averaged to obtain seasonal low-frequency estimates. In addition, estimates from the four seasons can be combined into annual estimates, the annual grid-point data within homogeneous climatic regions can be averaged, and these data can be filtered or averaged to obtain low-frequency annual estimates. All reconstructions should be examined to determine how much and which type of averaging is most appropriate. In this study both calibration and verification statistics were used to evaluate and assess the importance of the averaging on the precision of the results.

11. The reliability and significance of the filtered low-frequency data set could not be determined with certainty because *df* were inadequate for many significance tests. Yet the filtered low-frequency reconstructions appear to be accurate, accounting for 58 to 71 percent of the instrumental data variance on the average. The importance and reliability of low-frequency climatic variations should be examined carefully in future dendroclimatic work.

12. Clearly, temperature information can be extracted from spatial arrays of tree-ring chronologies from drought-sensitive sites, although the sign of the temperature relationship may be negative or positive and may vary from one season to the next. Collecting cores from trees growing on a drought-subjected site may emphasize the growth response to precipitation and drought, but this focus should not exclude study of the effects of other factors that have limited growth. Estimates of temperature from trees on arid sites may reflect a water-loss response as well as direct effects of temperature on growth. Temperature estimates from these arid sites must be carefully compared to temperature information available from high-altitude, upper tree-line and subarctic sites.

13. Sea-level pressure can be calibrated and reconstructed from tree-ring chronologies from arid sites even though pressure itself does not limit growth. The reconstruction is possible because sea-level pressure is correlated with temperature, moisture, wind, sunshine, and cloud cover, which are known to influence the environmental factors that limit growth. Other indices of atmospheric conditions such as the Southern and Northern Oscillation also can be reconstructed from spatial variations of tree-ring growth.

14. Variables of sea-level pressure and temperature, within certain limits, can be related to tree-ring variations at both nearby and distant sites, because large-scale features in these variables are evident. If only

relationships to climate in nearby sites are considered in an analysis of these variables, important large-scale climatic features may be lost. Variations in precipitation involve smaller-scale phenomena, and although large-scale features are present, they are less evident. A climate reconstruction effort should include an assessment of the time and spatial scales of the system, as well as the probable differences in the responses to seasonal climate. Calibration and verification procedures should evaluate all relevant time and spatial scales that can contribute significant variance.

15. The strategy of using the same data throughout the analysis should not necessarily be used in the future, unless there is a need to evaluate other procedures and methodology. When optimal grids of both tree-ring and climatic data are assembled, the irrelevant outliers, methodologies, and relationships should be discarded, but not before they are tested adequately. Too strict adherence to existing theory or to the reasonableness of a relationship may overlook important new possibilities, but if questionable relationships are considered, they should be scrutinized carefully, tested objectively, compared to other information, and accepted (at least conditionally) if they hold up in the evaluation. They should not be abandoned just because they are unpopular notions.

16. Many new data sets and techniques represent significant improvements over older ones. New tree-ring chronologies include more recent data and cover wider areas. Response surfaces created on personal computers include nonlinear relationships and interactions and make more attractive displays than response-function plots made on mainframe computers. ARMA techniques and the Kalman filter can handle some time-series relationships better than multiple regression. These new techniques do not invalidate all work accomplished with less powerful tools, however, and they inevitably have their own limitations, which will become evident as they are applied in a variety of situations. The older tools frequently can answer some of the same questions, perhaps not as elegantly, and usually their weaknesses and limitations are more apparent. Although scientific revolutions do occur, each generation cannot repeat all past experimentation just because new tools are available. Progress usually builds on, improves, and incorporates the results using old techniques rather than destroying and completely replacing them.

17. The results of this study are by no means final or conclusive. They represent a concerted effort to diagnose possible approaches to reconstructing spatial climatic features from spatial variations in ring-width chronologies. The techniques were the best available at the time, the analyses were based on current and proven dendrochronological and biological principles, and decisions were based on objectively derived

statistics with a sprinkling of human judgment. The large number of empirical approaches structured around canonical regression may be considered unsophisticated compared to some of the newer ways of handling spatial relationships. Nevertheless, it is hoped that this detailed description of each step in the analysis, along with discussions of the annual reconstructions, will encourage people to look more carefully at high-resolution dendroclimatic evidence for past climatic variability. The work demonstrates that spatial patterns of well-dated tree-ring series can be transformed into year-to-year estimates of some important climatic factors over a spatial grid. The consistency of the reconstructed precipitation, temperature, and sea-level pressure patterns with synoptic-scale climatic features suggests that a more dynamic understanding of past climatic variations may be obtained from analysis of tree-ring data than has been thought possible.

Appendix 1

Technical Notes

A. The Eigenvectors and Their Principal Components

Large numbers of tree-ring chronologies or climatic data can be reduced to a smaller number of orthogonal factors by extracting the most important eigenvectors. These eigenvectors are multiplied by the original data to obtain the corresponding principal components (PCs), or amplitudes. The procedure (Fritts 1971, 1976) begins with the calculation of the correlation matrix, $_mC_m$:

$$_mC_m = \left[\frac{1}{n}\right]_mF_nF'_m \tag{A.1}$$

where F is the set of m chronologies of index values for n years, and F' the transpose of F. All index values $_ix_j$ are normalized to obtain $_iF_j$ as follows:

$$_iF_j = \frac{_ix_j - \overline{m}_i}{sd_i} \tag{A.2}$$

where $\overline{m}_i$ is the chronology mean and sd_i is the standard deviation. A complete matrix of m eigenvectors is obtained, $_mE_m$, from the correlation matrix so that

$$_mC_mE_m = {}_mE_mL_m \tag{A.3}$$

where $_mL_m$ is a set of scalers called *eigenvalues*. These scalers have positive values in proportion to the variance reduced by each eigenvec-

tor and are used to arrange the eigenvectors in order of their importance.

Each column of $_mE_m$ is 1 eigenvector, with m elements. The first 15
columns were selected, representing the 15 eigenvectors that reduce the
most variance. The first 10 are mapped in Figure 4.1.

The PCs, or amplitudes (Fig. 4.2), were obtained by multiplying the
selected k eigenvectors by the normalized chronology data:

$$_nX_k = \,_nF_mE_k \tag{A.4}$$

B. Canonical Regression Using
Principal Components

All regression models used assume that the data are normally distributed and that the predictors are independent and uncorrelated
variables. Yet almost all time series of climatic or tree-ring data from a
geographical area are intercorrelated, that is, they are never completely
independent of other climatic or tree-ring series from the surrounding
region, because they all reflect the large-scale features of the macroclimate that are common to the data set. This intercorrelation creates one
or more common patterns or colinearity usually associated with one or
more of the most important eigenvectors of the data set and the
associated PCs.

Another portion of the variance may represent individual variability
or variance that is different from the variance common to the whole
data set. Some of this variability will be shared more with certain
chronologies than with others. This information may also be reflected in
the variations of the eigenvectors and their PCs. As the order of the
eigenvectors and their PCs increases, there is less associated variance,
the patterns they represent become smaller-scale, there is less shared
information among the elements of the original data set, and there is a
greater likelihood that more of the variations in the PCs arise by
chance, are errors in the data, or are unmodeled relationships.

If the chronology data are highly selected to represent the same
species growing on similar sites in a small geographic area, the trees are
likely to respond in a similar way to many of the variations in climate. A
large amount of the variance will be associated with a very few eigenvectors of growth. If the chronologies in a grid are for different species
or are sampled from sites over a wide spatial grid, however, a much
smaller amount of variance will be associated with any one eigenvector,

and a large number of eigenvectors and PCs may be needed to express the important modes of climatic response.

Multiple regression or other correlation techniques may underestimate the error and not provide valid significance estimates if large amounts of collinearity exist. Stepwise multiple regression, using collinear tree-ring chronologies as potential predictors of climate, enters individual chronologies as predictors while blocking other chronologies from being entered if they are highly correlated with predictors that are already included. A small number of chronologies are usually selected, along with the associated error terms, to estimate climate, and the information in the much larger unselected set of chronologies is left out.

Although chronologies can have large error components, particularly in the early years when the number of available trees may be small, every chronology in a semiarid region is usually well correlated with other chronologies in the same region (Fritts 1976). This variance in common represents a response to climate (Fritts 1976) and is potential information on climate. In addition, the colinear climatic data can be decomposed by extracting the eigenvectors of climatic data and calculating the PCs of climate. A number of PCs of the tree-ring chronologies can be related to a number of the PCs of climate by using canonical regression. The significance testing on the canonical variates, rather than on individual chronology or climatic data, provides an appropriate way of testing for broad-scale colinear features leading to reconstructions of climate. The approach used in this study was to select information that was meaningful from as large a number of chronologies as possible and to eliminate the information that was not meaningful by rejecting orthogonal variates that contributed little to the reconstruction of climate.

Contrary to the views of Stockton et al. (1985), more than a few eigenvectors are needed when using large spatial arrays of tree-ring data, and it is not necessary to be able to interpret the individual eigenvector patterns (although the first few eigenvectors often have interpretable features). The eigenvectors were used only to reduce the number of variables and to arrange the data into orthogonal modes of behavior. At the end of the analysis the reconstructed PCs were transformed back into the original real-world climatic estimates. A variety of criteria has been used to decide how many eigenvectors and PCs to employ in a particular analysis (Preisendorfer et al. 1981). The number of PCs selected must be large enough to represent the important modes of variance for the analysis but not so large that unnecessary *df* are lost (see Section G, below, for more discussion on selection of PCs).

C. The Calculations

Two computer packages, entitled TREPACK and PREPACK, were developed to perform the calibration and subsequent analysis. The PCs of the tree-ring chronologies were calculated for the time period beginning in 1601 and ending in 1963, and those for the climatic data were calculated from the period from 1901 or 1899 through 1970. The 15 (p) largest PCs from the tree-ring chronologies were calibrated with q PCs of climatic data using canonical regression analysis (Glahn 1968) on n years of dependent data from 1901 or 1899 through 1963 (Johnston 1972).

The PCs that have been derived from each data set give little or no information on the relationships between the two sets, however. This information is obtained by extracting the eigenvectors of the $p \times q$ correlation matrix and obtaining the canonical variates (Clark 1975).

The necessary statistical equation (Blasing 1978) is

$$_n(\hat{D} - \overline{D}c)_q = {}_n(C - \overline{C}c)_p[Scc]_p^{-1}A_qL_qB_q^{-1}[Scd]_q. \quad (A.5)$$

Matrix $\hat{D}$ contains q PCs of climate from which the means of the calibration period, $\overline{D}c$, have been subtracted. Matrix C contains the p largest PCs of the tree-ring chronologies, also corrected for the means of the calibration period, $\overline{C}c$. Matrix C can include PCs at two different lags or one set of PCs of data corrected for autocorrelation and another set of data uncorrected for autocorrelation (see Chap. 5, Sec. B). The terms Scc and Scd are the standard deviations used to normalize the data before generating the correlation matrix, which was in turn inverted for the canonical analysis. Matrices A and B are new orthogonal canonical variates where $q < p$ and L is a diagonal matrix of scaler coefficients called the *canonical correlations*. The canonical analysis rearranges and orders the correlations between and within C and D to form A and B, which represent statistically independent modes of correlation. Each q column of A is correlated with the same q column in B but is uncorrelated with the other canonical variates (Clark 1975). This canonical rotation restores orthogonality that was lost by forming couplet models with the PCs from the B, I, M, F, and FF data sets and by selecting subsets of years that were different from the years analyzed in the original PC analysis.

Since the canonical variates are orthogonal, the effect of any one canonical variate, along with its corresponding elements in A and B, can be eliminated by setting the corresponding canonical correlation to a zero value (Blasing 1978). Two different significance tests are available for canonical correlation. The most common one is the Wilks' Lambda Statistic, which tests the significance of each canonical correlation (Clark 1975). This test appears to be inappropriate for regression analysis, because it gives equal weight to the variance in both the predictand and predictor sets.

The second type of significance test uses the variance ratio between the explained and the unexplained predictand variance in the regression estimate. This test was calculated by subtracting the PCs of estimated climate, $\hat{D}$, from the PCs of climate, D, to obtain the residual matrix. This figure is squared by multiplication with its transpose and summed for all n years to form the residual sums of squares ($SSRES$) as follows:

$$SSRES = \sum_{k=1}^{n} \sum_{j=1}^{q} \left[_j(D - \hat{D})'_n (D - \hat{D})_j \right]. \qquad (A.6)$$

The sum of squares due to regression ($SSREG$) is calculated as:

$$SSREG = \sum_{j=1}^{q} \left[_j(\hat{D} - \overline{D}c)'_n (\hat{D} - \overline{D}c)_j \right]. \qquad (A.7)$$

The predictand variance reduced by each canonical variate is tested by entering one canonical variate at a time into regression, starting with the largest canonical correlation, and testing the variance reduced using an F statistic. The canonical correlations for all other variables not to be considered in that particular set are set to zero. The particular F value used for testing is

$$F = (SSINC/dfu)/(SSRES/dfr) \qquad (A.8)$$

where $SSINC$ is $SSREG$ at that step minus the variance in regression, $SSREG$, at the last step in which variable being tested was significant. The $SSREG$ is accumulated over all variates that are significant. The df for dfu is p, the number of tree-ring predictors, and the df for dfr is

$$dfr = (nq - pk)C \qquad (A.9)$$

where

$$C = \frac{1 - r_1}{1 + r_1} ; \qquad (A.10)$$

C is a correction of the number of df for the mean autocorrelation, r_1, of the residuals (Mitchell et al. 1966); n, q, and p are the same as above; and k is the number of nonzero canonical correlations at that step in the analysis. The correction, C, is applied only for $r_1 \geq 0$ to adjust for loss of df attributed to positive autocorrelation.

Whenever the F ratio is smaller than the theoretical value ($p > 0.95$), that particular canonical correlation is set to zero, and the F ratio for the next smaller variate is calculated again using only the canonical variates found to be significant at previous steps in the selection sequence. If the F ratio is equal to or greater than the expected value, that canonical variate remains in the regression. After a variable is selected and added to the regression, the testing procedure rechecks the variance ratios for variates that were previously rejected and enters them if the F ratio is now large enough to be significant. After all previously rejected variables have been retested and either selected or rejected, the stepwise procedure continues by testing the canonical variate with the next smaller canonical correlation.

This forward and backward selection process continues until all canonical variates have been tested. Sometimes variates with high canonical correlations are excluded from the regression. At other times they are excluded at an early stage but are included at a later stage because the variance of the residuals (the denominator in the F ratio) declines with the inclusion of more and more significant variates. The sum of the elements of the canonical variates retained in the final regression, after all insignificant ones have been eliminated, becomes the coefficient of the transfer function. Few variates with low canonical correlations proved to be significant.

The reconstructions for all independent data are obtained by applying the chronology PCs $_n(C - \overline{C}c)_p$ for n independent years to the significant canonical variates in Equation A.5. Matrix $\hat{D} - \overline{D}_c$ is then multiplied by the predictand eigenvectors, and the appropriate standard deviations and means are used to denormalize the result into $_n\hat{Y}_m$ estimates of $_nY_m$ climatic data. The equation may be written in terms of the original X tree-ring chronologies as

$$_n\hat{Y}_m = {}_n\left[\left([X - \overline{X}c]_k [Scx]_k^{-1} E_p - {}_n\overline{C}c \right)_p [Scc]_q^{-1} A_q L_q B_q^{-1} Scd_q + {}_n\overline{D}c_q \right]$$

$$\times \, F_m' Scy_m + {}_n\overline{Y}c_m \tag{A.11}$$

where $\overline{X}c\ Scx$, $\overline{Y}c$, and Scy are column means, and standard deviation of the x and y variables, E_p and F_m, the selected eigenvectors. Tree-ring chronology data can be inserted as X in this equation to obtain $_n\hat{Y}_m$, the climatic reconstructions.

The fractional explained variance, or calibrated variance, EV, is calculated as

$$EV = \sum_{j=1}^{q} \left\{ \left[\frac{_i(\hat{D} - \bar{D}c)'_n(\hat{D} - \bar{D}c)_j}{_j(D - Dc)'_n(D - Dc)_j} \right] V_j \right\} \quad (\text{A.12})$$

where V_j is the proportion of climatic data variance reduced by the jth eigenvector. EV measures the amount of variance reduced by that particular calibration. EV is also calculated for each climatic station. The fractional EV values are multiplied by 100 to convert them to percentage of calibrated variance.

The similarities between the estimated and actual yearly climatic patterns also may be assessed for each year by calculating the correlation and the reduction of error between the estimates and actual data. The reconstructions from Equation A.11 are first normalized by

$$_n\hat{P}_m = {}_n(\hat{Y} - \bar{Y}c)_m(Scy^{-1})_m \quad (\text{A.13})$$

where $_n\bar{Y}_c$ and Scy are the means and standard deviations for the m stations over n years of the calibration period and $_n\hat{P}_m$ denotes the normalized estimates.

The correlation coefficient for year i, r_i, is

$$r^i = \frac{_i(P - \bar{Pr})_m(\hat{P} - \bar{\hat{P}}r)'_i}{\sqrt{_i(P - \bar{Pr})_m(P - \bar{Pr})'_i} \sqrt{_i(\hat{P} - \bar{\hat{P}}r)_m(\hat{P} - \bar{\hat{P}}r)'_i}} \quad (\text{A.14})$$

where $\bar{\hat{P}}r$ denotes the means of $\hat{P}$ climatic estimates, and $\bar{Pr}$ denotes the means of the climatic data for year i. The reduction of error (RE) for year i is

$$RE_i = 1 - \frac{_i(P - \hat{P})_m(P - \hat{P})'_i}{_i(P - \bar{Pr})_m(P - \bar{Pr})'_i} . \quad (\text{A.15})$$

These statistics were calculated in the same manner for each grid point over all years (Eqs. A.14 and A.15). They, along with the percentage of variance calibrated, were used to evaluate each calibration.

The explained variance in Equation A.12 can be converted to a population estimate using the available df in the residuals and the equations described by Kutzbach and Guetter (1980). The adjusted statistic, EV', is calculated from the EV statistic as follows:

$$EV' = [EV(nqC - q) - pk]/(nqC - pk - q) \quad (A.16)$$

where n is the number of years used for calibration, p is the number of predictors, q is the number of predictands, k is the number of canonical variates included in regression, and C is the correction defined in Equation A.10. An early version of this equation also multiplied the pk values by C. This earlier version was used in estimating the EV' described in this volume.

D. Verification Statistics

1. Correlation Coefficient

The Product Moment Correlation reflects the entire spectrum of variation that is common between two data sets, including both high and low frequencies. Correlations calculated from the first differences, r_d, measure only the high-frequency variations expressed by the year-to-year differences.

The data sets are assumed to be normally distributed, so the values of r and r_d are tested by comparing them to the 0.95 probability limit for a zero correlation calculated as

$$r^* = \frac{t^*}{\sqrt{n - 2 + t^{*2}}} \quad (A.17)$$

where t^* is the one-sided 0.95 level for the t statistic with $n - k \; df$ where $k = 2$ for r and $k = 3$ for r_d (Panofsky and Brier 1968).

A significant correlation is one that is greater than r^* and one implying that the variance in the two data sets is linearly related. However, the correlation does not imply that the values in one data set lie near those in the other data set, nor does it imply that the scale of the variations, the amount of variance, in the two sets is similar. Other tests are needed to evaluate these types of differences.

2. Sign Test

A nonparametric and less sensitive measurement of reliability is to count the number of times that the signs of the departures from the sample means agree or disagree. Inferences can be made about the probabilities of two dichotomous outcomes: (1) that the estimate $\hat{Y}_i$ and the observation Y_i are on the same side of the dependent data mean and (2) that they are not. In the verification test the estimate can be considered a "success" with outcome 1 and a "failure" with outcome 2.

The number of positive signs expected by chance follows a binomial distribution and is $1/2n$, where n is the total number of observations. For $n < 45$, the cumulative distribution tables for the binomial distribution with parameter $p = 0.5$ (Beyer 1968) can be used to evaluate the critical value ($p = 0.95$). For $n > 45$, the binomial distribution is well approximated by the normal distribution. If the test is successful, it can be concluded that the sign of the estimate is more often correct than would be expected from random numbers. The test measures the associations at all frequencies but does not assume that the data are normally distributed.

A similar test is made for the first differences. This test is similar to the one above but reflects the high-frequency climatic variations.

3. Product Means Test

The product means (PM) test (Fritts 1976) was originally proposed by Terence J. Blasing and is not described in the statistical literature. It has intuitive appeal as a diagnostic tool because it is straightforward, and it accounts for both the signs and the magnitudes of the similarities in two data sets. It emphasizes the larger deviations from the mean over the smaller ones.

The PM test calculates the products of the deviations and collects the positive and negative products in two separate groups based on their signs. The values of the products in each group are summed and the means computed. The difference between the absolute values of the two means $M_+ - |M_-|$ is tested for significance.

If the estimates bear no relationship to the observed values, the positive and negative products will occur with about equal frequencies, and the absolute values of the expected means will be the same with a zero difference. If there is a real relationship, the positive products will be more numerous, the mean of the positive products will be larger than the mean of the negative products, and the difference will be large and positive.

The t statistic for the difference of two means (Dunn and Clark 1974) is used for testing. The statistic used can be expressed as

$$t = \frac{M_+ - |M_-|}{\sqrt{\dfrac{\sigma_+^2}{n_+} + \dfrac{\sigma_-^2}{n_-}}} \tag{A.18}$$

where $|M_-|$ is the absolute value of the negative mean, n_+ and n_- are the number of positive and negative products, and σ_+^2 and σ_-^2 are the corresponding variances. The value of t, computed from Equation A.18, can be compared to a critical value of t^* obtained from a table of percentage points for the t-distribution (Beyer 1968). For sample sizes greater than 30, the table is entered with $n_+ + n_- - 2$ df at the one-sided 0.95 probability level. For sample sizes smaller than 30, an adjustment is made in the df used (Bickel and Doksum 1977).

The PM test, unlike the majority of the verification tests, is not a standard statistic with well-known characteristics. Therefore, Gordon and LeDuc (1981) use simulations to evaluate the PM statistic, and they report that this test is much more likely to underestimate than to overestimate the value of the true relationship, especially when the relationship is a weak one. This result apparently occurs because the distributions of the observations are truncated at zero and do not approximate a normal distribution. The error terms are overestimated, so that the PM tests fail more often than would be expected from normally distributed data. The statistic was retained as one of the verification tests in spite of this weakness, because it uses the product to discriminate between the magnitude as well as the sign of the agreement. Its use would not jeopardize the verification procedure by passing more than 5 percent of the insignificant relationships. Some relationships that were truly significant may have been rejected, however.

4. Reduction of Error

The reduction-of-error statistic (RE) provides a sensitive measure of reliability, and it is similar in some respects, but not equivalent to, the explained variance statistic for the calibration (Lorenz 1956, 1977). The equation used to calculate the RE statistic (Eq. A.15) can be simplified where $\hat{Y}_i$ estimates and the Y_i independent data are departures from the dependent period mean value:

$$RE = 1.0 - \frac{\sum_{1}^{n}(\hat{Y}_i - Y_i)^2}{\sum_{1}^{n}Y_i^2}. \tag{A.19}$$

The term on the right of Equation A.19 is the ratio of the total squared error obtained with the regression estimates and the total squared error obtained using the dependent period mean as the only estimate (Lorenz 1956, 1977; Kutzbach and Guetter 1980). This average estimation becomes a standard against which the regression estimation is compared. If the reconstruction does a better job at estimating the independent data than the average of the dependent period, the total error of the regression estimates is less, the ratio is less than one, and the *RE* statistic is positive.

The *RE* can be partitioned into three component parts (Gordon and LeDuc 1981; Gordon 1980) that can be used to analyze sources of error affecting a particular climatic reconstruction. The equations for the components are

$$\sum_1^n \hat{Y}_i Y_i = n\bar{\hat{Y}}\bar{Y} + (n-1)\widehat{Cov}(\hat{Y}, Y) \tag{A.20}$$

and

$$RE = -\frac{\sum_1^n \hat{Y}_i^2}{\sum_1^n Y_i^2} + \frac{2n\bar{\hat{Y}}\bar{Y}}{\sum_1^n Y_i^2} + \frac{2(n-1)\widehat{Cov}(\hat{Y}, Y)}{\sum_1^n Y_i^2} \tag{A.21}$$

where

$$RE = RISK + BIAS + COVAR. \tag{A.22}$$

As in equation A.19 the calculations are based on departures from the dependent period mean except for $\widehat{COV}(\hat{Y}, Y)$. $\widehat{COV}(\hat{Y}, Y)$ is

$$\sum_1^n \left(\hat{Y}_i - \bar{\hat{Y}}\right)\left(\bar{Y}_i - \bar{Y}\right)$$

and $\bar{\hat{Y}}$ and $\bar{Y}$ are the sample means from the independent period.

The *RISK* term is always negative, and its absolute magnitude is a comparative measure of the variability of both the estimates and the actual observations used in testing. The *RISK* term is a lower limit for *RE*, below which the regression reconstructions will exhibit no skill at all in reproducing the variations in the instrumental data (Gordon 1980). Ideally, a model should produce as much variance in $\hat{Y}_i$ as in Y_i so that the *RISK* = -1.0.

To obtain a positive *RE*, the *RISK* term must be offset by the accuracy of the estimates, as indicated by the second and third terms, representing bias and covariation. The *BIAS* is positive when the mean of the estimates is on the same side of the calibration mean as the actual independent climatic data used for the verification testing. It is negative when the estimated mean is on the opposite side of the calibration mean compared to the mean of the instrumental data. Component *COVAR* reflects the strength of the correlation between $\hat{Y}_i$ and Y_i and measures the similarity of the temporal patterns in the estimates and observations.

If the *RE* is negative but its value is greater than the *RISK* term, the reconstructions may still contain some meaningful climatic information. A negative *RE* is an undesirable statistic, however, and indicates that improvements are needed to make the estimates at least as good as using only the mean of the dependent data for an independent estimate.

Gordon and LeDuc (1981) show that the *RE* statistic does estimate the explained variance fairly accurately, but they also report that the size of the sample being analyzed, n, appears to affect the *RE*'s significance. For $n = 20$, an *RE* statistic of zero is roughly equivalent to a 0.95 confidence limit, but for $n = 10$ or 15, an *RE* statistic would have to be greater than 0.25 or 0.12, respectively, to be significantly different from zero at the 0.95 confidence limit.

5. Contingency Analysis
and the Chi-Square Statistic

A contingency analysis (Beyer 1968) involving a table with m rows and columns was designed to test for a relationship without making any assumption about linearity. The value of m should be small enough to ensure that there are, on the average, five or more observations per cell. A chi-square statistic is then used to test whether the relationship is sufficiently strong to be significant. A significant test ($p = 0.95$) implies that the estimated and the observed climate are related in some way, although neither the type of association nor the direction of the relationship is assumed in the testing. Linear, nonlinear, or inverse relationships can be detected by examining the distribution of values in the contingency table. If the chi-square is significant, a contingency coefficient can be calculated to estimate the degree of association in the contingency table in a manner analogous to the correlation coefficient.

The chi-square test used in the contingency analysis was applied to the data pooled over all stations. A variety of other statistics was calculated from the pooled data to reveal any major deviations from a normal distribution. Two of these were the ratios of the skewness and kurtosis, where the divisor is the skewness and kurtosis of the instrumental data, and the numerator is the skewness and kurtosis of the reconstructed data.

6. Spatial Correlation

Livezey and Chen (1983) provide guidelines for assessing the collective significance of a finite set of intercorrelated statistics. Following these guidelines, we can derive the critical significant percentages of grid points from the binomial distribution as follows: for the 96 precipitation or sea-level pressure grids, it is 9.4 percent of the total number of points, and for the 77 temperature grids, it is 10.1 percent. To assess the effects of spatial correlation within the data sets, one thousand Monte Carlo simulations of the averaging and differencing procedures were performed for each variable and for both the instrumental and the reconstructed data, using randomly chosen extreme years. The respective threshold percentages of grid points needed for significance of the collective set of statistics were (1) 17.5 and 1.5 for the pressure instrumental data and reconstructions, (2) 15.5 and 0.0 for the temperature instrumental data and reconstructions, and (3) 27.5 and 0.0 for the precipitation instrumental data and reconstructions.

7. Additional Verification Statistics

The reconstructions also were analyzed by year by calculating the correlation coefficient, reduction of error, the sign test, and the *PM* statistic to the grid-point estimates. These results were expressed as the percentage of years passing each test. In addition, the *RE* statistic was calculated and pooled over all stations (j) in the entire grid and for all stations in each climatic region within the grid as follows:

$$RE_p = 1 - \frac{\sum\limits_{j=1}^{m} \sum\limits_{i=1}^{n_j} \left(\hat{Y}_{ij} - \hat{Y}_{ij}\right)^2}{\sum\limits_{j=1}^{m} \sum\limits_{i=1}^{n_j} Y_{ij}^2}. \qquad (A.23)$$

A value of RE_p was obtained for a region by substituting the number of stations in the region for m in Equation A.23. To minimize the effect of extreme values, the pooled *RE* statistic was calculated as:

$$RE_p^+ = 1 - \frac{\sum\limits_{j=1}^{m} \sum\limits_{i=1}^{n_j} \left(Y_{ij} - \hat{Y}_{ij}d_j\right)^2}{\sum\limits_{j=1}^{m} \sum\limits_{i=1}^{n_j} Y_{ij}^2} \qquad (A.24)$$

where

$$d_j = 1 \quad \text{when station } j \text{ had } RE > 0$$

$$d_j = 0 \quad \text{when station } j \text{ had } RE \le 0.$$

For stations with a negative *RE*, Equation A.24, in effect, replaces the model estimates with the estimate of a zero deviation but still reflects the variance unaccounted for by stations with a positive *RE* statistic. A positive value of RE_p^+ but not of RE_p can be interpreted to mean that there is still reliable information in many of the reconstructions despite the fact that the total variance of the estimates is not larger than zero.

E. An Example

The following example illustrates the process of canonical variate selection and verification using model 16T15I15M to reconstruct winter temperature for the 77-station grid (Table A.1). The largest (highest-ranking) 16 PCs of winter temperature were estimated from the largest 15 PCs of the chronologies uncorrected for autocorrelation and the largest 15 PCs of the chronologies corrected for first-order correlation, making a total of 30 potential predictors of climate.

The first step in the process was the calculation of all canonical correlations and their corresponding coefficients. Selection began (Table A.1) by entering the first canonical pair of variates into regression, which was the pair that had the highest correlation, and calculating the sums of squares of the variance reduced (Eq. A.7). The *F* ratio (Eq. A.8) for the pair was 0.57, a nonsignificant value, so the first canonical variate was rejected. In steps 2 and 3 the second and third most highly correlated pairs were tested in like manner. The *F* ratio for the third variate pair exceeded the 0.95 significance level of 1.51, and that canonical variate was selected. The first-order autocorrelation of the residuals at step 3 was 0.185, which was a higher value than was

Table A.1. An Example of Statistics Obtained at Different Steps in the Canonical Regression Selection Procedure

Step Number	Canonical Variates		Statistics of Residuals				% Variance			Corrected Variance Calibrated
	Entered for Testing	Already in Regression	r_1	nq minus pk[a]	dfr	Reduced by Entered Variable[b]	Remaining in Residuals[c]	Reduced by All Included Variables	F Ratio	
1	1		0.148	978	725	3.58	6.32	3.58	0.57	2.2
2	2		0.138	978	740	6.28	6.07	6.28	1.03	3.7
3	3		0.185	978	672	23.75	5.91	23.75	4.02[d]	14.2
4	1	3	0.157	948	690	3.58	5.61	27.33	0.64	16.4
5	2	3	0.145	948	708	6.28	5.34	30.03	1.18	18.0
6	4	3	0.169	948	673	8.07	5.54	31.82	1.46	19.1
7	5	3	0.179	948	659	6.44	5.74	30.19	1.12	18.2
8	6	3	0.178	948	661	9.90	5.56	33.65	1.78[d]	20.3
9	1	3, 6	0.150	918	678	3.58	5.27	37.23	0.68	22.5
10	2	3, 6	0.134	918	701	6.28	4.98	39.93	1.26	24.0
11[e]	4	3, 6	0.161	918	663	8.07	5.18	41.72	1.56[d]	25.1
16	8	3, 4, 6	0.156	888	648	8.62	4.90	50.34	1.76[d]	30.4
21	9	3, 4, 6, 8	0.154	858	629	11.45	4.51	61.79	2.54[d]	37.2
23	2	3, 4, 6, 8, 9	0.097	828	681	6.28	3.88	68.07	1.62[d]	40.9
25	5	2, 3, 4, 6, 8, 9	0.087	798	670	6.44	3.66	74.52	1.76[d]	44.9

Note: Regression on winter temperature using the 77-temperature grid and model 16T15I15M, which reduced 44.9% of the temperature variance.

[a] dfr before adjusting for autocorrelation.
[b] $SSING / dfu$.
[c] $SSRES / dfr$.
[d] Significant, as it exceeded the chosen critical value of 1.51.
[e] Only data for steps with significant F ratios are shown in subsequent rows.

obtained in steps 1 and 2. This result reduced the available *df* of the residuals (*dfr*) from 740 to 672, yet the *F* ratio was still sufficiently high to allow the variable to be entered. The variance reduced by regression was 23.75 percent, a figure that amounted to 14.2 percent of the total variance of the temperature data.

Steps 4 and 5 retested canonical variates 1 and 2, but neither combination reduced significant variance. Steps 6 through 8 tested canonical variate pairs 4, 5, and 6. Only the variance reduced by entering canonical 6 was significant. It amounted to 9.90 percent, which was substantially less than that reduced at step 3, and the *F* ratio was barely significant. The autocorrelation of the residuals was almost as large as it was in step 3: 0.178 (Table A.1). The adjusted *df* of the residuals was now 661, because two canonical pairs were entered into regression. Together they reduced 33.65 percent of the PC variance, which amounted to 20.3 percent of the total winter temperature variance. Variables 3 and 6 were used in steps 9 through 11 to retest variates that previously had been insignificant. The residual variance was reduced sufficiently to pass variate 4 at step 11 with an *F* ratio of 1.56, which was barely significant. The three variables reduced 41.72 percent of the PC variance and 25.1 percent of the temperature variance.

The testing procedure continued without a significant improvement in regression until step 16, when variate pair 8 proved to be significant, with 50.34 percent of the PC variance and 30.4 percent of the winter temperature variance calibrated. After the selection procedure retested canonical variates 1, 2, 5, and 7, pair number 9 was the next one to be significant. It was entered at step 21 with 61.79 percent of the PC variance and 37.2 percent of the winter temperature variance calibrated. The residual variance and its autocorrelation had become sufficiently reduced by step 23 that canonical variate 2 was now entered as significant. At this time, the first-order autocorrelation of the residuals declined to 0.097, which caused the residual *df* to increase to 681. The residual variance declined from 4.51 to 3.88, and the corresponding *F* ratio rose to 2.54. At step 23, 68.07 percent of the PC variance was calibrated, which included variable 9 as a significant variable in regression. The last variate pair to pass the significance test was 5 at step 25. No additional canonical variates were found to be significant after step 25, and the canonical selection procedure was terminated. The *F* ratio at step 25 was 1.76, and the final seven-variable canonical regression calibrated 74.52 percent of the PC variance; when the variance unexplained by the temperature eigenvectors was added, the calibration percentage was reduced to 44.9 percent of the winter temperature

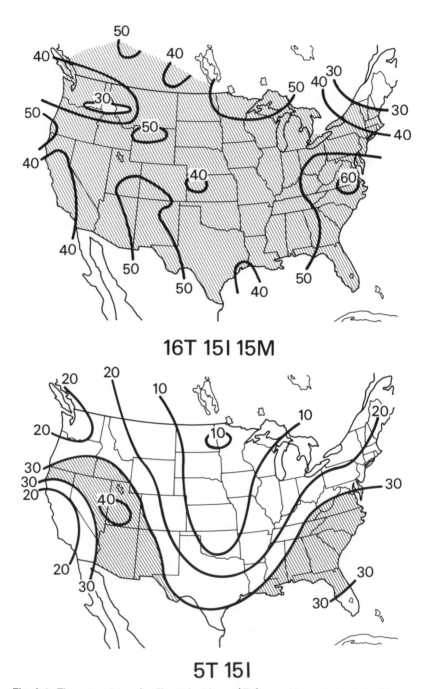

Fig. A.1. The percentage of calibrated variance (*EV*) over the reconstructed grid for models 16T15I15M and 5T15I for winter temperature. *Shaded areas*: variance greater than 30 percent.

variance. It should be noted that the latter, more conservative, percentage was reported throughout this study.

The rank of the canonical correlation and the Wilks' Lambda Statistic do not necessarily indicate the importance of the canonical variate to the regression. Canonical variate pair 3 with a mean square of 23.8 reduced the most variance (Table A.1). Variate pair 9 was next in importance, and variate pair 2 was the least important. Canonical variate pair 1, which had the largest canonical correlation, apparently accounted for too little predictand variance to be chosen for the transfer function.

When Cook et al. (1988) compared canonical regression to principal component regression, they did not include this F test. Cook assumes that his test of the partial correlation gave an equivalent result (Cook, personal communication 1988). If a test of this difference had been made, it could well have improved the canonical regression statistics over those of the principal component regression results. Thus, the canonical regression could be superior rather than just appearing to be equivalent to the methods used by Cook et al. (1988). It is a disappointment that the study of Cook et al. (1988) was not an open experiment, so that this weakness could have been discussed and identified before, rather than after, their work had been completed. Such an approach could have made their test a more fair and more objective analysis.

Calibration percentages exceeded 50 percent for stations in the southeastern United States, the southern Rocky Mountains, and the southern High Plains (Fig. A.1). Eureka, California; Yellowstone, Wyoming; Fargo, North Dakota; Marquette, Wisconsin; and Edmonton, Alberta, Canada, also exhibited values exceeding 50 percent. Areas with less than 40 percent calibrated variance were southern California, the Columbia River Basin, and New England. Individual stations with less than 40 percent variance reduced were Regina, Alberta; Hays, Kansas; and Galveston, Texas. It is interesting that the highest variance reduced was 63 percent for the Lynchburg, Virginia, record. It is likely that high values at such a distant station may be a chance result. This possibility was checked by applying verification statistics to independent temperature data from that record, and these data indicated only marginal significance at best. This example illustrates the critical importance of the verification testing. Verification statistics appeared to be more reliable than calibration statistics for model-selection purposes.

The reconstructions and instrumental data could well be correlated for a small number of years and uncorrelated for the remaining years. This possibility was evaluated (Table A.2) using the correlation (r) and RE statistics for each i year (Eqs. A.14 and A.15) calculated over the data from the 77 stations. The correlation coefficients were zero or

Table A.2. Number of Years for which Correlation Coefficients (r) and Reduction-of-Error Statistics (RE) were Larger than or Equal to Selected Values and the Calibrated Variance for a High-Ranking Component, Merged, and Annual Models

Statistic	Level	Single Component Calibration	Merged Model for Winter	Annual Average
r	0	60	58	60
r	0.3	51	47	55
RE	0	38	37	46
RE	0.5	16	10	11
RE	0.6	9	6	5
		Variance (%) Calibrated[a]		
Unadjusted		44.9	41.6	47.7
Adjusted[b]		27.2	24.7	37.0

Note: Out of 62 years for winter, 61 years were used for annual average. Table is based on models 16T15I15M and 16T15I15M + 5T15I for winter temperature, 77-grid.
[a] R^2 is used instead of EV for all averaged sets (see text).
[b] Adjusted using Eqs. A.15 and A.16 (see text).

greater for all but 2 years. They equaled or exceeded values of 0.3 fifty-one times. The more stringent reduction-of-error statistic was zero or greater for 38 years and exceeded values of 0.5 for 16 out of 62 winters. The 44.9 percent calibrated variance, the wide distribution of percentages greater than 30 over the grid (Fig. A.1), and the good distribution of statistics in different winters indicated that climatic patterns appeared to be well reconstructed by this model for a large geographic area and for a large number of years.

The variance reduced and df from step 25 (Table A.1) were entered in Equations A.10 and A.16 to make the necessary corrections for the explained variance adjustment.

$$C = (1 - 0.087)/(1 + 0.087) = 0.840 \qquad (A.25)$$

$$EV' = \left[0.449(63 \times 16 \times 0.840 - 16) - 30 \times 7\right] /$$

$$(63 \times 16 \times 0.840 - 30 \times 7 - 16) = 0.263 \quad (A.26)$$

The earlier version's calculation of EV' for this model was 0.301.

The estimated RE' of Kutzbach and Guetter (1980) was also calculated for the calibration period, as follows:

$$RE' = EV' - (pk/[nqC + q])(1 - EV') \qquad (A.27)$$

The solution of Equation A.27 for the example above is 0.084. It was 0.157 for the earlier version.

In an attempt to estimate the variance at low frequencies, the EV' and RE' were also calculated for data after they had been treated with the low-pass filter, passing 50 percent of the variance at periods of 8 years (Fritts 1976):

$$EV'_f = [EV(n'/8) - 2]/[(n'/8) - 3] \qquad (A.28)$$

where n' is the available df and the divisor of eight is a conservative estimate of the number of years between relatively independent observations. Two df are subtracted from the numerator, one for the mean and one for the end effects of the filter. One more degree of freedom is subtracted from the denominator for the correlation coefficient. The term n' is calculated as:

$$n' = (nqC - pk - q)/q \qquad (A.29)$$

where n, q, p, k, and C are the same as above. As in Equations A.16 and A.27, an early version of the equation multiplied pk by C (Equation A.10), and these estimates were used in the analysis. The value for EV'_f is substituted for EV' in Equation A.27 to obtain the RE'_f for the filtered value.

The fractional calibrated variance for the population is 0.263, substantially less than the original EV of 0.449 obtained for this particular example. The seven canonical variates each consumed 30 df, reducing the available df of the residuals to 670. For this reason, it is important to make these corrections so that the results are a realistic appraisal of the "true" calibration. It should be stressed that the more conservative EV' was used in making evaluations.

Using Equation A.29, we see that only 620 df remain after the statistics are discounted in the equations above. The df remaining for each station reconstruction after canonical regression, n', was obtained by dividing this value by 16. Thus, 39 df at each station or PC out of a total of 63 (years) remained for statistical testing. The earlier version estimated 41 df available after calibration. The later version was more conservative and preferable, but the differences were not considered to be sufficiently important to justify recalculation of all model calibrations.

Table A.3. Examples of Verification Statistics for High-Ranking Component Models and Selected Merged Model

Model	% Tests Passed[a]							% Statistics > 0			% Years Passing	
	Correlation	Correlation First Differences	Sign Test	Sign Test First Differences	Product Means Test	All Tests	Correlation	Correlation, First Differences	Reduction of Error	Sign Test	Product Means Test
5T15I	43	46	22	28	26	33	93	87	46	9	17
16T15I15M	50	46	13	19	33	32	87	91	35	13	18
5I + 16IM	52	56	24	20	39	38	93	91	57	13	21

Note: Based on winter temperature, 77-station grid.
[a]Five statistics were calculated at 54 stations, making a total of 270 tests.

Verification results (Table A.3) for winter temperature models 16T15I15M and 5T15I are presented with a merged reconstruction of the two models, denoted as 5I + 16IM. These reconstructions were simply the arithmetic averages of the grid points for the two reconstructions.

Usually, the verification tests for the leading models greatly exceeded the levels expected solely by chance, even when the values were adjusted for the effects of spatial correlation (Livezey and Chen 1983). The percentage of all tests that passed significance in Table A.3 ranged from 32 to 38. They were well above the 15.5 percent expected by chance for temperature. Some statistics were lower, however. In the case of model 16T15I15M, only 13 percent of the sign tests were significant. This result suggests that the 16T15I15M model generates reconstructions in which the signs of the departures from the mean were not well estimated. Other statistics, such as the *PM* test, clarify the results further. They indicate that the larger deviations were more correctly estimated by the 16T15I15M model and that incorrect signs appear to occur when the deviations from the average were small. The overwhelming majority of the *RE* and correlation statistics were positive for all three models, and a great many of the correlation tests were significant. These data imply that both reconstructions reproduced the temporal patterns of climate variability substantially better than would be attributable to chance variations.

There were few individual years with enough observations to make very meaningful verification tests over space on a yearly basis. Only 20 years had more than thirty stations included over the entire grid; the average was eleven stations per year. Less than one-half of the grid was

Table A.4. Examples of Verification Statistics for Pooled Station Data from High-Ranking Component Models and the Selected Merged Result

Model	Chi-Square[a]	Coefficient of Contingency[b]	Ratio of Skewness
5T15I	124.34	0.291	10.333
16T15I15M	143.11	0.311	0.527
5I + 16IM	146.83	0.314	1.642

Note: Based on winter temperature, 77-station grid.
[a] The 0.95 critical level is 83.39.
[b] The maximum value is 0.943.

Table A.5. Examples of Reduction-of-Error Statistics for
High-Ranking Component Models and the Selected Merged Result

Model	EV' Fractional Explained Variance	RE_p Pooled Reduction of Error	RISK	BIAS	COVAR	RE_p^+ Pooled Positive Reduction of Error
5T15I	0.223	−0.152	−0.337	0.011	0.174	0.077
16T15I15M	0.449	−0.334	−0.923	0.210	0.379	0.143
5I + 16IM	0.416	−0.090	−0.478	0.111	0.277	0.123

Note: Based on winter temperature, 77-station grid.

represented, and most of those stations were in the eastern half of the United States. Still, 9 and 17 percent of the years passed the sign and *PM* tests, respectively. In general, the poor spatial coverage and the nonuniform distribution of the independent temperature data did limit the usefulness of verification tests on a yearly basis.

The contingency analysis of the pooled data (Table A.4) did not differentiate among these reconstructions. All of the chi-square statistics were significant, being well beyond the 0.95 critical level of 83.39. The skewness ratio suggests that the 5T15I reconstructions may underestimate the frequencies of low temperatures in the independent data. The 16T15I15M reconstructions overestimate the same feature. Note that the combined model resolves these two conflicting traits, resulting in an improved estimation of the skew in the independent data. This kind of result was frequent and supported the strategy of combining the reconstructions from two or three models of different structure.

All three models had negative pooled *RE* statistics (Table A.5), yet the *RE* statistics reported for the stations were positive in one-third to one-half of the cases (Table A.3, col. 10). The high, pooled, positive reduction-of-error statistic, RE_p^+, and the high *COVAR* term indicated that even though the 16T15I15M reconstruction had the most negative *RE* statistic, it had the highest covariation with the independent data. This result implies that selected stations were reconstructed exceptionally well by the 16T15I15M model but that other stations were reconstructed poorly. The lower *RISK* term, the higher RE_p, and the higher number of stations with positive *RE* values for the 5T15I reconstructions indicate that this reconstruction had a lower relative variance but appeared to be a better overall reconstruction.

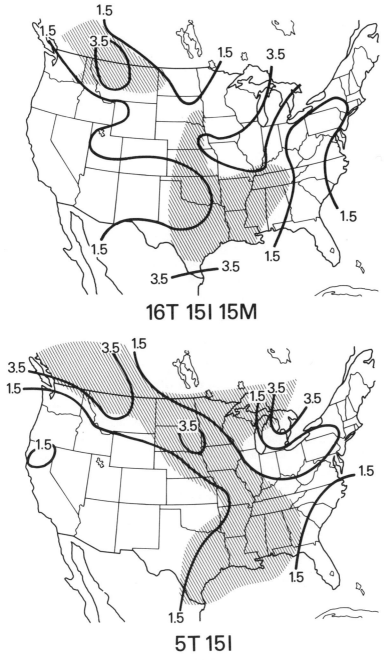

16T 15I 15M

5T 15I

Fig. A.2. The numbers of verification tests (out of a total of 5) that were significant for models 16T15I15M and 5T15I for winter temperatures. *Shaded area*: grid points with positive *RE* statistics.

The regions with the best verification statistics were in the area from the western Canadian provinces and throughout the northern Great Plains to the Gulf Coast (Fig. A.2). In this example, neither the numbers of tests passing nor the *RE* statistics were acceptable along the United States Pacific Coast, in the Southwest, or along the southeastern coast. It should be pointed out here that many stations in the West had insufficient data to make valid verification tests (see Figs. 3.2 and 3.3). The reconstructions differed in the amount of verification in the Great Lakes region, where 5T15I failed more significance tests than 16T15I15M, but there were more *RE*s with positive values in this area.

Note that verification in both models was generally better for areas to the east of the tree sites, including the central Plains, prairie states, and Great Lakes, than for many stations next to and within the grid of tree sites. The spatial patterns of calibrated variance were not necessarily the same as the spatial patterns in the verification statistics. These results could represent important differences in the reconstructions, but some of the variance may have arisen from errors in the earliest instrumental data. These possibilities can be evaluated to some extent by calibrating models of different structure and noting the stations with and without significant verification statistics.

F. Merging

1. Program VAREX

Program VAREX was developed to average two or more reconstructions and to calculate new statistics for the averaged estimates by comparing them to the appropriate climatic data used for calibration and verification. It was not possible to compute the *EV* by Equation A.12, though, because the PCs of $\hat{D}$ and D were not available. The pooled R^2 was obtained as a substitute:

$$R_{pooled}^2 = \frac{\left[\sum_{j=1}^{m} {}_j(P - \bar{P}c)_n'(\hat{P} - \hat{P}c)_j \right]^2}{\left[\sum_{j=1}^{m} {}_j(P - \bar{P}c)_n'(P - \bar{P}c)_j \right]\left[\sum_{j=1}^{m} {}_j(\hat{P} - \bar{\hat{P}}c)_n'(\hat{P} - \bar{\hat{P}}c)_j \right]}$$

$$(A.30)$$

where $\hat{P}$ and P are the estimated and actual normalized data (Eq. A.13), n is the number of years, and m is the number of stations.

Although the R^2 statistic was not identical to EV, there appeared to be no other alternative. This statistic can be adjusted in the same way as EV by applying Equations A.16 and A.25–A.29 using the pooled df of the individual model calibrations. In the following text the fractional variance calibrated by the merged models is multiplied by 100 to obtain a percentage.

2. Examples of Merging

Some of the merged model statistics (winter average) were compared to the average of the two best component models in Table A.2. The numbers of correlation coefficients and RE statistics above selected threshold values, as well as the calibrated variances for the merged model, were generally below those of the calibration model averages, reflecting the fact that the variance was reduced whenever reconstructions were averaged. It is necessary to distinguish between simple changes in variance due to averaging and changes that arise from the relative improvement in the agreement between the instrumental record and the reconstructions. One method is to compare the explained variance of the combined model with the averages of the individual model estimates. If the explained variance of the combined model is greater than the average of the individual models, the difference is more likely to reflect improvement.

Other statistics for the merged model are included in Tables A.3–A.5. With the exception of the sign test of first differences, the combined reconstructions exhibited a greater percentage of stations passing each verification test. The greatest improvement was seen in the percentage of stations with a positive RE (Table A.3) and the larger (more positive) RE_p value (Table A.5). The chi-square, the coefficient of contingency, and the ratio of skewness (Table A.4) all showed improvement. These verification statistics indicated that both reconstructions contain useful climate information and suggest that the combination had fewer and perhaps smaller errors. The pooled verification results generally showed more improvements than the individual model averages, but considerable variability can be observed in different statistics.

While the percentage of variance calibrated by the merged model (Fig. 6.4) approximated the averages of the individual models mapped in Figures A.1 and A.2, significant verification tests (Fig. 6.4) were distributed over a wider area indicating that in the merged reconstructions (Fig. 6.4) the verification was as good as or better, including areas in the Pacific Northwest, the Plains states, and eastward to the Great Lakes and central Gulf Coast. As was the case for the individual reconstructions, verification was not significant for stations in Oregon, southern Idaho, northern Wyoming, and the Southwest.

G. Further Model Development

1. Varying the Number of Principal Components

Two basic strategies were used to determine how many and which PCs of climate should be calibrated. The first was to determine, in some quantitative way, which eigenvectors were below the noise levels and could be eliminated without affecting the reconstruction. The second was to construct and calibrate models with different numbers of predictand PCs and use the resulting calibration and verification statistics to select the optimum combination.

In the first strategy several different methods were used to identify the noise-level PCs. Method A was simply to use all the eigenvectors necessary to reduce 90 percent of the variance of a given variable. This value ranged from a minimum of 9 eigenvectors for winter temperature to a maximum of 37 for summer precipitation and 39 for the 65 ring-width chronologies. In method B the average value for all nonzero eigenvectors was calculated, and the number of cases with eigenvalues greater than the average value was determined. These numbers ranged from 5 for winter temperature over western North America to 23 for summer precipitation. The number was 15 for the 65 ring-width chronologies, and that is the number used for all temperature and precipitation reconstructions.

Other objective methods were (C) the "scree" line approach of Tatsuoka (1974) and (D) the simulation approach of Preisendorfer and Barnett (1977). Method C used the values of the eigenvalues plotted as a function of their order. All eigenvectors were selected that had eigenvalues above the point where the change in slope became nearly constant, and those below that point were excluded. Method D involved generating one hundred grids of random data approximating the numbers and lengths of the dependent data (i.e., 77×70 for the temperature stations), calculating the correlation coefficients among the grid points one hundred times, extracting the eigenvalues, and then using these results to obtain $+2$ standard deviation limits for each rank of the eigenvalues. The eigenvector numbers with eigenvalues above the lower limit were selected. Methods C and D provided more stringent selection criteria ranging from 3 to 12 eigenvectors, depending on the variable and the grid.

The extreme variability in the numbers of significant eigenvectors, both within and among methods, confirmed that there is no simple answer as to how many eigenvectors are important. It seemed best to start with a fairly high number of eigenvectors based on A and B of the first strategy and then to use the most objective test possible to eliminate error variance at a later stage in the analysis. The use of the F statistic to limit the number of canonical variates entering the regression seemed to be a better solution than arbitrarily excluding PCs without any evaluation of a statistically significant relationship.

2. Varying Model Size and Structure

The following statistics were calculated for the calibration period:

1. calibrated variance percentage,
2. the number of years with $r > 0$,
3. the number of years with $r > 0.3$,
4. the number of years with $RE > 0$,
5. the number of years with $RE > 0.09$,
6. the number of years with $RE > 0.5$, and
7. the number of years with $RE > 0.6$.

The earliest studies used only these calibration statistics to select the optimum models. When the complete program for verifying the independent data became operational, the following verification statistics were calculated using the available independent climatic data:

8. the number and percentage of stations over the grid with $r > 0$,
9. the number and percentage of stations over the grid with r significant,
10. the number and percentage of stations over the grid with r of the first difference > 0,
11. the number and percentage of stations over the grid with r of the first difference significant,
12. the number and percentage of stations over the grid with $RE > 0$,
13. the number and percentage of stations over the grid with $RE >$ significance level for r,
14. the number and percentage of stations over the grid with sign test significant,
15. the number and percentage of stations over the grid with sign test of first difference significant,
16. the number and percentage of stations over the grid with product means test significant,
17. the number and percentage of stations over the grid passing one test,

Table A.6. Examples of Selected Statistics from Two Singlet Structures and Eleven Couplet Structures

1 Model	Dependent Data			Independent Data						
	2 % Variance Calibrated	3 No. of Canonical Variates Included	4 % Years RE > 0	5 % Significance Tests Passed	6 % RE > 0	7 Coefficient of Contingency	8 % Years Passing Sign Test	9 Product Means Test	10 % Stations Passing > 2 Tests	11 Model Rank
15T15B15I[a]	29.8	3	0.476	16	21	0.308	17.5	25.0	21	2
15T15MB15I[a]	29.9	4	0.452	21	23	0.277	12.5	18.8	32	5
15T15B15M[a]	31.6	4	0.500	12	26	0.289	11.3	22.5	13	5
15T15MB15M[a]	18.0	1	0.355	6	9	0.307	7.5	20.0	4	12
15T15I[b]	22.2	3	0.435	24	34	0.269	15.0	18.8	34	4
15T15I15M[a]	42.1	6	0.597	30	26	0.342	15.0	27.5	51	1
15T15I15F[a]	16.7	1	0.419	11	26	0.334	11.3	21.3	11	9
15T15M15F[a]	33.2	4	0.565	6	11	0.263	8.8	15.0	8	10
15T15I15MF[a]	24.8	2	0.387	21	34	0.293	13.8	21.3	34	3
15T15M15MF[a]	28.0	3	0.452	12	8	0.285	8.8	23.8	17	8
15T15F[b]	14.7	3	0.355	9	8	0.288	7.5	13.8	15	13
15T15F15MF[a]	27.7	3	0.565	17	21	0.263	10.0	20.0	23	7
15T15F15FF[a]	18.1	2	0.426	5	8	0.290	8.8	22.5	6	11

Note: Based on winter temperature, 77-station grid.
[a] Couplet structure.
[b] Singlet structure.

18. the number and percentage of stations over the grid passing two tests,
19. the number and percentage of stations over the grid passing three tests,
20. the total number and percentage of tests passed,
21. the number of sign tests passed by year over all stations,
22. the number of product means tests passed by year over all stations,
23. the chi-square value, and
24. the similarity of reconstruction to a normal distribution as compared to the similarity of the instrumental data set.

The seven calibration and the seventeen verification statistics were tabulated for the eleven basic couplets and the two singlet models for each variable, season, and grid (Table A.6). Each statistic shown in columns 2 and 4–10 in the table was ranked, the ranks summed by row, and the sums ranked again in column 11 to identify which basic model had the best overall potential. The four highest-ranking models, shown in column 11, were 15T15I15M, 15T15B15I, 15T15I15MF, and 15T15I.

These four highest-ranking model structures were selected, and 2 to 20 temperature PCs were calibrated and verified for each one. This process produced 76 new calibrations and verifications for each variable, season, and grid. The calibration and verification statistics were tabulated and the models ranked to help with the selection. In the example using winter temperature (Table A.6), the results from the first-ranking model and the fourth-ranking model structures (shown in column 11) were selected. The fourth-ranking model was selected because it has superior verification statistics and a simpler structure than the second- and third-ranking models.

The IM and I model structures (Figs. A.3 and A.4) provide good examples of the variations encountered when different numbers of PCs are used in the model development procedure. The IM models (Fig. A.3) with the largest number of climate PCs (shown on the right) seemed to have the highest calibrated variance. However, the coefficient of contingency was relatively high for models with 3, 4, 5, and 7 PCs. The number of *RE* statistics greater than zero generally rises with greater numbers of PCs, but this number can vary substantially, depending on which canonical pairs are selected. In this example, models with 4, 10, 16, and 18 PCs had the most verification tests significant. Those for 4, 10, and 16 PCs were chosen for subsequent merging. Model 18, although similar to 16, was excluded, because it had a slightly lower calibrated variance, a lower number of verification tests passing, and the same number of canonical variates in regression.

T IM

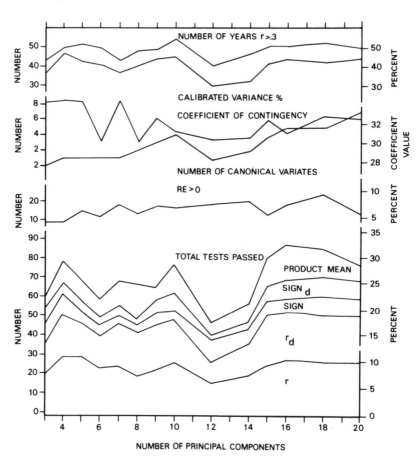

Fig. A.3. Selected statistics of couplet *IM* model structures for 77-grid winter temperature (the first-ranking model in Table A.6) using 2 to 20 PCs of climate. *Plots 1 and 2*: calibration statistics (listed as 3 and 1 in App. 1, Sec. G.2); *Plots 3 and 4*: the coefficient of contingency and the number of canonical variates that were significant; *remaining plots*: verification statistics (listed as 12, 20, 16, 15, 14, 11, and 9 in App. 1, Sec. G.2). Except for the number of canonical variates, rising values correspond to an improvement.

The I models (Fig. A.4) showed a more typical pattern of rising statistics followed by declining statistics with increasing numbers of PCs. This pattern was most evident in the verification statistics. In this example, two minima at 8 and 14 PCs interrupted the downward pattern that began with the entry of 5 PCs into the canonical regression. Models with 5, 7, and 14 PCs were chosen for subsequent merging and analysis.

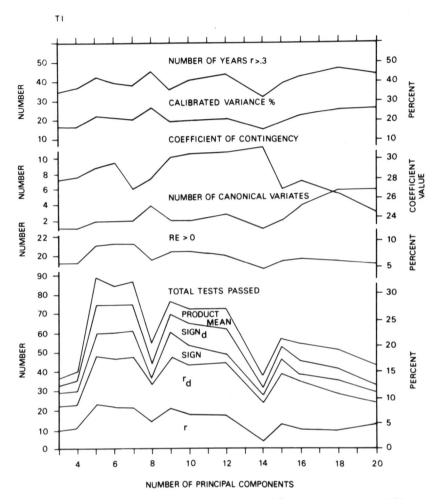

Fig. A.4. Same as Figure A.3, for the singlet *I* model (the fourth-ranking model in Table A.6).

Reconstructions from models selected in the aforementioned fashion were then averaged into combinations of two and three models of varying structure. Approximately twenty to thirty combinations were considered for each variable, season, and grid. The calibration and verification statistics were recalculated, and selected results were plotted for final model selection. At early stages in the work sometimes more than three models were averaged (Fritts, Lofgren, and Gordon 1979), but improvement in the verification statistics rarely occurred in these cases, so it was decided to combine no more than three models for any one variable, grid, or season.

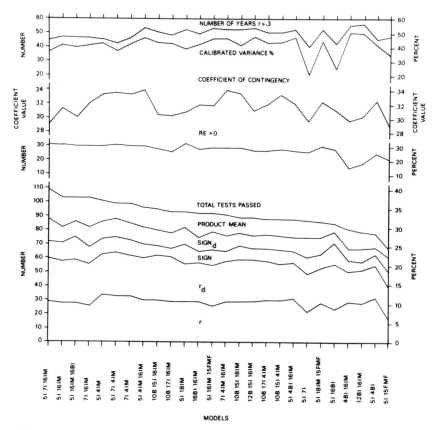

Fig. A.5. Selected statistics (Fig. A.3) of 26 high-ranking model combinations for winter temperature, 77-grid, arranged from highest to lowest number of verification tests that were significant (see App. 1 for model name notation).

The statistics shown in Figure A.5 are from the twenty-six leading model combinations constructed and tested for winter temperature for the 77-station grid. The models are ordered from highest to lowest number of verification statistics that passed the significance testing. The statistics are the same as those shown in Figures A.3 and A.4, except for the number of canonical variates, which was not applicable to the merged reconstructions. The highest-ranking models, shown on the left, included examples from the I singlets and BI, IM, and F(M)F couplets.

Some statistics for eight of the highest-ranking models shown in Figure A.5 are included in Table A.7, along with the statistics for the component models that were included in them. The notation, such as 5I + 7I + 16IM (Table A.7), was simplified as 5I 7I 16IM in Figure

A.5. This particular merged model had the highest number of significant verification tests passing but was fourth-ranking when all statistics were considered. Model 5I + 16IM had a higher calibrated variance, a higher coefficient of contingency, and was first-ranking. Model 5I + 4IM + 16IM had the highest calibrated variance and coefficient of contingency of any model in the table, but thirteen fewer verification tests were significant, and it was only fifth-ranking. Model 5I + 16IM not only was highest-ranking, but it appeared to be a good selection, as it had only six fewer verification statistics significant than Model 5I + 7I + 16IM, the same number of *RE* statistics greater than zero, and the fourth-ranking calibration percentage.

In almost all seasonal merged results a number of promising models had very similar statistics. The following rationale was used to break ties. The simplest models were thought to have the least reconstruction bias and therefore to have the most reliable climatic information. If the statistics were approximately equal, it seemed prudent to select the simpler model. That was defined as the one with the fewest singlets or couplets and with the highest-ranking merged components; if the number of singlet and couplet components was equal, the one with components made up of the fewest predictand PCs was chosen.

For example, the merged model 5I + 16IM for winter temperature (as shown in Table A.7) was the average of only two components (the 5I singlet and the 16IM couplet). It was selected over the more complex 5I + 7I + 4IM model. In addition, it was the average of the 5I and 16IM components, which ranked first and third among the components. The alternative model contained two I models and a 4IM model, which ranked fourth. The selection was supported by the fact that the latter model, which had been excluded, had four fewer significant verification tests and 2.7 percent less variance calibrated.

The reconstructions from the best combinations for each of the four seasons were weighted by the number of months in the season, and the weighted seasonal values were averaged for temperature and pressure or totaled for precipitation to obtain the annual values. In the case of precipitation, the estimates were expressed as a percentage of the 1901–1970 amounts. The year began with December and ended with November to conform with the chosen seasons. The instrumental data for seasons were combined in the same way and used to calculate the necessary statistics.

3. Developing Models for Sea-Level Pressure

Two issues of concern arose: Should the correlation or covariance matrix be used in forming the PCs of pressure? Could the procedures

Table A.7. Examples of Selected Statistics for Highest-Ranking Merged Models and Their Components

| Merged Models | % Variance Calibrated | No. of Tests Passed Out of 371 | RE > 0 | Pooled RE | | | | No. of Regions with Positive RE | No. of Regions with Positive Pooled RE | Coefficient of Contingency | Rank |
				All Values	Positive Values Only	Negative Values Set to Zero	Pooled Positive RE				
5I + 7I + 16IM	37.0	109	31	−0.089	0.174	0.118	0.474	7	4	0.293	4
5I + 16IM	41.6	103	31	−0.096	0.181	0.123	0.459	7	4	0.314	1
5I + 16IM + 16BI	40.0	103	30	−0.079	0.178	0.117	0.427	8	5	0.302	3
7I + 16IM	41.3	103	30	−0.117	0.170	0.114	0.412	7	3	0.321	6
5I + 4IM	42.9	101	30	−0.085	0.177	0.120	0.398	7	4	0.335	2
5I + 7I + 4IM	37.9	99	31	−0.068	0.184	0.116	0.472	7	4	0.337	1
7I + 4IM	42.7	99	30	−0.102	0.166	0.113	0.374	7	4	0.335	7
5I + 4IM + 16IM	46.9	96	30	−0.115	0.168	0.114	0.353	7	4	0.340	5
Single Models											
16BI	19.3	60	22	−0.218	0.107	0.045	0.903	8	3	0.298	5
5I	22.3	89	25	−0.155	0.145	0.077	0.855	7	4	0.291	1
7I	20.6	88	26	−0.170	0.128	0.069	0.726	7	3	0.269	2
4IM	47.6	78	9	−0.423	0.212	0.024	0.435	4	0	0.345	4
16IM	44.9	87	19	−0.350	0.143	0.053	0.331	7	2	0.311	3

Note: Based on winter temperature, 77-station grid.

used for temperature and precipitation be simplified for pressure to reduce the number of variables that had to be calibrated and verified? These issues were resolved in the following manner.

a. Correlation or Covariance Matrix

The correlation rather than the covariance matrix had been used to calculate the PCs of temperature and precipitation, because it was thought that the large differences in variance among the temperature and precipitation stations would contribute more error than signal to this particular analysis. In earlier calibrations of sea-level pressure (Kutzbach 1970; Fritts et al. 1971; Blasing 1975) the covariance matrix had been used, because both the variance and the correlations among the grid-point values were considered important to the dynamics of the spatial pressure patterns that were to be examined. Thus, the use of the covariance matrix was justified on the basis that it portrayed a more meaningful representation of PCs of surface pressure than the correlation matrix.

It had never been determined experimentally, however, if there was any real difference in the calibrations with tree-ring information using correlation versus covariance. A comparison was made by calibrating five models of different structure (15P15B15I, 15P15I, 15P15I15F, 15P15IM15F, and 15P15F) for both winter and spring pressure by first using the correlation matrix and then using the covariance matrix. The correlation method reduced about 7 to 8 percentage points more calibrated variance than the covariance method, but the verification statistics exhibited no corresponding improvement. In fact, the *RE* statistics were usually lower for the correlation method. Therefore, the small difference between the two methods did not warrant changing this particular procedure, so that covariance matrix was used for pressure, as was done earlier.

b. Sea-Level Pressure Calibration

The problem of simplifying the sea-level pressure calibration was divided into two subproblems: (1) determining the maximum number of predictands and (2) determining the maximum number of predictors for the transfer-function equation.

(1) The maximum number of predictands was determined in the following manner. Given the fact that the reconstruction of pressure from ring-width chronologies relies on the interrelationships between the microclimates of the tree sites (estimated in our case by temperature and precipitation measurements) and the general circulation of the

Table A.8. Counts of Significant Verification Tests for Three Leading Component Models and the Merged Model

Model	No. of PCs Tested	% Tests Passed						% Statistics > 0		
		Correlation	Correlation, First Differences	Sign Test	Sign Test, First Differences	Product Means Test	All Tests	Correlation	Correlation, First Differences	Reduction of Error
11P12I	11	36	27	18	18	45	29	100	73	27
11P10B10I	11	55	18	36	18	36	33	82	64	18
15P15B15M	15	40	33	20	27	47	33	80	60	0
I + BI + BM	11	45	45	36	27	55	42	91	91	27

Note: Based on winter sea-level pressure.

atmosphere measured by surface pressure, it was hypothesized that the tree rings would not reconstruct any more detail in sea-level pressure than could be reconstructed from the temperature and precipitation measurements from the area of the tree-ring grid.

To evaluate this hypothesis, the number of PCs of seasonal pressure was varied from 9 to 15 and calibrated with the non-noise-level PCs for both temperature and precipitation. As was the case for calibration of temperature and precipitation, the calibration statistics of pressure usually increased with an increase in the number of predictand PCs, but the verification statistics reached maximum values when there were approximately 11 pressure PCs. Thus, the maximum number of PCs used to calibrate sea-level pressure for each season was inferred to be 11, although somewhat larger numbers sometimes were considered, and a few were found to provide good verification.

(2) The maximum number of predictors was set to 15 for a "singlet" model and 30 for a "couplet" model. The model types with the best statistics for winter sea-level pressure used 15 PCs of climate and included components 15B15M, 15B15I, and 15I as the best predictors. The 11 PCs of pressure were then calibrated with 7 to 14 PCs of the tree-ring chronologies. In the case of the BM model, 4 to 14 PCs of the tree-ring chronologies were calibrated.

4. Examples Using Sea-Level Pressure

An example of the sea-level pressure analyses appears in Table A.8. It uses the three winter pressure models (11P12I, 11P10B10I, and 15P15B15M) and the average of these models, designated as model I + BI + BM. This combination had the superior statistics.

The verification tests used in the model evaluations were applied to the pressure PCs developed from the independent data. Table A.8 summarizes the verification statistics for the three calibrations and the merged result. Model 11P12I had the smallest number of significant statistical tests, but all of the correlation coefficients were positive and the *RE* statistics were superior to those of the other two models. Models 11P10B10I and 15P15B15M had 33 percent of the verification tests significant, but the latter had no *RE* statistics that were positive. In all cases, the numbers of significant statistics were well above the 5 percent expected by chance variations. The results for the merged reconstructions appeared to be an improvement; however, there were fewer *RE* statistics with positive values than there were for winter temperature, yet the numbers of tests passing were comparable (Table A.3).

Table A.9. Reduction of Error for the First Five Independent PCs from Three Leading Component Models and the Merged Model

| PC | Model | Reduction of Error and Related Statistics | | | | | No. of Tests Passed |
		RE	RISK	BIAS	COVAR	Correlation	
1	a 11P12I	−0.180	−0.501	0.010	0.312	0.223	4
	b 11P10B10I	−0.357	−0.797	0.010	0.430	0.243	4
	c 15P15B15M	−0.727	−1.362	0.014	0.621	0.268	3
	I + BI + BM	−0.135	−0.600	0.011	0.454	0.297	5
2	a 11P12I	0.015	−0.138	−0.007	0.161	0.219	2
	b 11P10B10I	−0.474	−0.511	−0.005	0.041	0.029	0
	c 15P15B15M	−1.655	−1.883	0.012	0.217	0.079	2
	I + BI + BM	−0.314	−0.454	0.000	0.139	0.104	1
3	a 11P12I	−0.059	−0.172	0.006	0.107	0.131	3
	b 11P10B10I	−0.407	−0.231	0.008	−0.184	−0.195	0
	c 15P15B15M	−2.020	−2.558	0.047	0.490	0.162	1
	I + BI + BM	−0.205	−0.364	0.020	0.138	0.123	1
4	a 11P12I	−0.490	−0.535	−0.029	0.075	0.054	0
	b 11P10B10I	−0.035	−0.458	−0.012	0.436	0.326	2
	c 15P15B15M	−1.335	−1.574	0.007	0.231	0.092	2
	I + BI + BM	−0.181	−0.417	0.011	0.247	0.194	1
5	a 11P12I	0.094	−0.403	0.027	0.470	0.378	3
	b 11P10B10I	−0.049	−0.511	0.093	0.368	0.296	2
	c 15P15B15M	−0.838	−2.314	0.085	1.391	0.471	3
	I + BI + BM	0.023	−0.789	0.065	0.743	0.440	4

Note: Based on winter sea-level pressure.

Table A.9 includes five statistics and the number of verification tests that are significant for the first 5 PCs of winter pressure reconstructed from the three models and the averaged result. Only a few *RE* statistics were positive, as the *COVAR* term was not large enough to overcome the *RISK*, and the *BIAS* term contributed little variance. Thus, the errors in these particular models for winter were as large as or larger than the average variance reduced by the dependent period average. It should be noted, however, that very few reconstructed PCs failed

Table A.10. Pooled Reduction of Error, Correlation, and Explained
Variance for Leading Component Models and the Merged Model

| Model | RE_p Pooled Reduction-of-Error Statistics[a] | | | | | EV' Fractional Explained Variance | |
	RE	RISK	BIAS	COVAR	Pooled Correlation[a]	Average of Sub-samples[b]	Full Calibration[c]
11P12I[d]	−0.125	−0.353	0.000	0.228	0.193	0.228	0.169
11P10B10I[d]	−0.294	−0.565	0.000	0.271	0.177	0.289	0.287
15P15B15M[e]	−1.101	−1.650	0.000	0.549	0.210	0.499	0.390
I + BI + BM	−0.153	−0.507	0.011	0.343	0.245	—	0.376

Note: Based on winter sea-level pressure.
[a]Pooled statistics were calculated from unnormalized eigenvector PCs.
[b]Average fractional explained variance for the six subsample calibration, not available for the merged model.
[c]Explained variance for the final calibration, using the entire 1899–1963 calibration data set.
[d]11 PCs were used in the pooled calculations.
[e]15 PCs were used in the pooled calculations.

totally, that is, with an *RE* statistic that was less than the *RISK* term. Virtually all of the reconstructed PCs had positive correlations with the instrumental data. The relatively high *COVAR* term indicated that at least the temporal patterns were correctly estimated. More importantly, the number of significant verification tests was comparable to that for winter temperature, and the correlation coefficients were large for the first and most important PCs of sea-level pressure.

The *RISK* terms for the first 3 PCs (Table A.9) were highest for the 11P12I model with the fewest predictors and lowest for the 15P15B15M model with the most predictors. It appears that overspecification with large numbers of predictors may have introduced excessive amounts of variability. For example, the *RISK* for PC 3 was −2.558, which indicates that the variance of the reconstructed PC was two and a half times larger than the variance in the instrumental measurement PCs.

Table A.10 summarizes the statistics obtained after pooling all of the independent PCs. The fractional explained variance for the six subperiods and for the full calibration using the entire 65-year period is also included in the table. The components of the *RE* showed variations similar to those noted in Table A.9. Model 15P15B15M had the largest *COVAR* term of the three, but the *RISK* indicated that the variance was overspecified by 165 percent. The merged model appeared to provide an acceptable compromise among the first three models, as the RE_p, *RISK*, and *COVAR* components were average, the pooled correlation was higher, and the fractional explained variance was nearly as high as in model 11P10B10I. The difference between the merged model and its components was not as large as it was for winter temperature; 1.4 percent less variance was calibrated, but 9 percent more of the verification statistics was significant. Both the calibration and verification statistics of the seasonal models improved substantially when they were combined into average annual values.

Appendix 2

Important Limitations

Some weaknesses and limitations inherent in this study are acknowledged in the following pages. Procedures that minimize the effects of certain limitations are also listed.

1. Only a small portion of the variance in a series of ring-width measurements from a single tree is potentially a "signal" reflecting past variations in climate. The nonclimatic variance, or "noise," from a single width measurement is often much larger, and it produces large uncertainties if only one width measurement is used to reconstruct climate. A variety of dendrochronological practices and procedures was used to minimize the noise and enhance the proportion of signal to noise (SN ratio) in the ring-width data sets. They include (a) sampling ten to twenty-five old climate-limited trees of a given species from one specific site type, (b) preparing the samples to reveal the ring structures clearly, (c) cross-dating the rings to ensure their correct placement in time, and (d) standardizing the measurements to remove any trends attributable to increasing age and changes at the site. Standardization converts the nonstationary ring-width series to stationary time series that can be more readily related to climatic variation. The SN ratio of the individual cores is greatly enhanced by averaging the standardized data from many trees to obtain a site chronology. In addition, the principal components that are entered into the transfer function and canonical analyses minimize the error and small-scale relationships while they maximize the variance in the large-scale relationships.

2. It is acknowledged that some climatic trends near the beginning of the record might resemble some of the estimated growth trends and hence some climatic information might have been lost because of the standardization process. A negative exponential curve with the maximum slope at the beginning of the series was usually fit to the ring-width measurements. If a curve of that shape did not fit the data, a straight

line or polynomial curve was attempted and tested. In most cases, the exponential curve was an adequate model. The ring-width measurements were then divided by the values of the fitted curve to obtain an index. Only the spectral estimates near infinity, time scales around 200 years that are greater than half the life span of the sampled trees, differ substantially from the red-noise variance spectra of most annually averaged climatic data sets. This result suggests that variations for time scales of 200 or more years, if they were present in the tree-ring measurements, were eliminated by the standardization process.

3. The quality of tree-ring chronologies varies greatly. In this study only the highest-quality chronologies were selected. Chronologies with undesirable statistical characteristics and obvious abnormalities, including long-term growth trends unrelated to tree aging, were eliminated from consideration. The individual chronologies were not tested for all violations of underlying assumptions of statistical tests (Graumlich and Brubaker 1986). Instead, the effects of occasional outliers and other non-normalities in individual chronologies were minimized by selecting chronologies with high-quality statistics, combining the chronologies, and extracting the most important eigenvectors and their PCs. Tests of normality were made on the calibration and verification sets, and no large pattern of non-normality could be detected. It was concluded that the use of such a large data set did minimize problems with occasional outliers and other non-normalities in individual chronologies.

4. The reconstructions for a particular variable, season, and grid point have large errors of estimate. According to the Central Limit Theorem, if a series of reconstructions is normally distributed and behaves like random numbers, the expected value for N values would be the arithmetic mean of the data and the variance of that mean would decrease with increasing sample size (Eq. 2.1). If the reconstructions also contain a signal, which in this case is yearly variation that is common to data points in a particular climatic region, the estimates of that signal have errors that are less than the random number case. The averaged data have less variance than the individual estimates, but usually the errors are also reduced so that the SN ratio increases. The error of any statistic, calculated from a reconstructed data set, must be evaluated carefully in terms of the amount of averaging used to obtain the climatic estimates. For example, a large change in the average temperature for the year may appear inconsequential when it is compared to variance of daily temperatures, and the change often will be smaller than the error of the daily measurements. The true significance of the change will become apparent if it is compared to the variance of annual averages rather than the daily measurements.

5. Errors appear to be present in some of the temperature and precipitation data, from the nineteenth and early twentieth centuries, and sizable errors have been noted around the periphery of the sea-level pressure grid. The errors in the tree-ring data were greatest at the beginning of the record, in which the numbers of trees sampled were the least. Large index values had larger errors than small values.

6. Many types of errors can be propagated through the transfer function. They would have been most severe for couplet models using many PCs, especially if large positive amplitudes had been averaged with large negative amplitudes to obtain a particular reconstruction. Errors may also be large for climatic stations at a great distance from the tree sites, because the strength of the linkages between the climate at two localities diminishes with increasing separation distance. Errors are greater for climatic conditions reconstructed from large, as opposed to small, chronology indices. In general, warm dry conditions reconstructed from low growth in arid sites are likely to have smaller errors than cool moist conditions. Many errors are reduced, but not eliminated, by averaging several calibrations and by combining the seasonal estimates into annual values. Care was taken in this study to recognize where relationships may be subject to large errors in the analysis.

7. The least-squares process of optimization virtually ensures that the model will be more accurate for the calibration period than for any other time period (Gordon 1982). The calibration and reconstruction processes assume that some fundamental predictor-predictand relationships are present in both the dependent data used for calibration and the independent data used for reconstruction (Gordon 1982). Any difference in the degree and kind of relationships between these two data sets will affect the reliability of the independent estimates. Both the tree-ring and climatic data were examined carefully before the analysis, and no data exhibiting obvious changes between the calibration and verification periods were used in this analysis. Verification statistics were emphasized in the model selection procedure to ensure that the calibrated relationships were preserved in the independent estimates.

8. Regression analyses cannot always discriminate among colinear relationships to identify exactly which features represent temperature, which represent precipitation, and which represent sea-level pressure estimates. Although objective tests have been applied in a carefully designed analysis to diagnose the feasibility of the methodology and to deduce joint patterns in climate, the result is a compromise and, at best, only a first approximation of the real-world system, as indicated by large error terms and residual variance.

9. A large number of model structures were calibrated and tested, and some tests on independent data were used to evaluate the statistical significance of each model result. Only those models with independent statistics (Livezey and Chen 1983) well above levels expected by chance were considered for further model development, and usually there were a number of models of different structure from which to choose. The same verification tests, along with other dependent data tests, were used to rank and to decide which of the verified models had the most reliable reconstructions. Marginal models that passed the initial screening were eliminated by this ranking. Though the models that were constructed included lags and leads between growth and climate that were physiologically or statistically reasonable, the approach used in this study was basically empirical, so that the results were subject to chance variations. The text and Appendix 1 describe the variety of objective procedures used to minimize some of the effects of chance variations, but these practices did not remove the random effects entirely. The reconstructed spatial fields described in this book are the best transformations that could be made with current technology from these arid-site tree-ring data for 1602–1963. Until other comparable data on climatic variations over western North America can be generated, these empirical reconstructions, including all of their errors, at least provide objectively determined, self-consistent estimates of annual climate variations for a large area of North America and the adjacent North Pacific based on high-quality tree-ring data.

10. Methodological questions were examined, the best procedures available at the time were adopted, and the same procedures were used throughout the analysis. A number of technological developments and data sets became available too late to be used without repeating large portions of the analysis. The most important were ARMA modeling, some types of spatial modeling, revisions in the sea-level pressure made by Trenberth and Paolino (1980), and new tree-ring collections from both the eastern and western United States.

11. Because of the diagnostic nature of this work, the climatic data grid was tested as a whole. All of the climatic stations or grid points, originally selected for calibration and verification, were retained throughout the analysis, even those data points that could not be reconstructed. The testing strategy did not allow for deletion of grid points to optimize the climatic estimates. The problem of poor quality was minimized, in some cases, by averaging the data or by confining attention to the areas where a large number of the reconstructions were statistically significant, but it was not eliminated.

12. Continuity of climate was assumed between variables plotted over space as points on a grid. Climatic variables that were correlated

over space and time were related to tree-ring chronologies over the same space and time. The statistical transfer functions that were obtained do not prove cause-and-effect relationships, but they assume that some underlying cause-and-effect linkages exist. It is recognized that trêe growth may not be directly affected by atmospheric pressure variations, but it can be assumed that the pressure fields are linked to temperature, moisture, wind, clouds, and sunshine, which are known to influence tree growth. The linkages involve the movement of cyclones, anticyclones, and fronts in the atmospheric circulation as they affect limiting conditions to growth over the sampled grid.

13. The relationships that led to calibration and reconstruction were approximated in this study by linear mathematical expressions (Lofgren and Hunt 1982). It was assumed that the averaging of many chronologies and climatic stations to obtain the principal components and canonical regression reduced most nonlinear features of the individual chronologies to more or less linear forms. It seemed more important to consider lagging relationships between years, autocorrelation, and differences between seasons than to consume *df* modeling weak nonlinear relationships that may or may not exist. Nonlinear effects were ignored in this particular study of large-scale dendroclimatic relationships. The same strategy is probably less appropriate for analysis of a single chronology from a particular species and climatic variations in the same area.

14. Arid-site tree-ring chronologies often are correlated with somewhat different climatic factors, depending on the species, site, and season. It was assumed that these differences in correlation represented real differences in the growth response that could be used in a multivariate calibration with coefficients of different weights to portray different chronology relationships. The initial calibrations reconstructed temperature, precipitation, or sea-level pressure one season at a time from the chronology variance. In these seasonal models the errors of estimate were relatively large because the anomalous growth effects of climate in the other three seasons were not assessed and statistically held constant. Even in such cases, some models had statistics that were well above noise levels. When the reconstructions for the best seasonal models were combined into annual values, the errors that had been noted in the seasonal reconstructions were substantially reduced. It was not easy to assess the significance of these types of interrelationships, but the flexible analysis of seasonal features appeared to produce meaningful annual results.

15. The error terms and the *df* of time series may be affected by autoregressive and moving average processes (Jenkins and Watts 1968) and by correlations that exist over space (Livezey and Chen 1983). This

study deals with this problem by estimating *df* losses due to autoregression and spatial correlation. Only the autocorrelation at lag one was used for the initial prewhitening of the tree-ring data. The PCs were used in part to minimize the spatial correlations and to simplify the calculation of the estimates. The data were not fully ARMA modeled. After the final model structures had been determined, fully ARMA-modeled series were substituted for all M series and the reconstructions were recalculated. The values and numbers of verification statistics using the new series showed no apparent overall improvement. Thus, the reconstructions using the M series were used throughout the entire analysis. This process was not a fair test of the ARMA method. It should have been used at the very beginning if it had been available when the basic structures of the calibration models were under development, and it probably would have improved the estimates even if it had not improved the statistics.

16. The reconstructions at the margins of the climatic grid may have larger errors than those in the interior, because few good chronologies were available from that region. In addition, there may be features in the circulation unique to the Pacific Coast which were not captured by the reconstructions because so many of the tree-ring sites were responding to the climate east of the Sierra Nevada and the Cascades, which serve as a major climatic barrier (Bryson and Hare 1974). This problem may be resolved for future reconstructions, because many new tree-ring collections have been developed for the western coastal region and in the eastern United States. Nevertheless, the discrepancy between glacial advances during the Little Ice Age and the warm temperatures reconstructed for California, western Oregon, and western Washington could result from errors of this nature. Also, it should be noted that the verification of annual sea-level pressure in the area of the Aleutian Low was poor (Fig 6.3). This inconsistency indicates uncertainty as to how well the circulation features in this area were reconstructed for the Little Ice Age, although the averages for all three centuries have considerably smaller errors than the individual annual estimates shown in Figure 6.3. The climate for large areas in the eastern United States could not be reconstructed because the climate stations were too far from the tree-ring chronology grid.

Appendix 3

User's Manual for Program DIFMAP

By Xuemei Shao and Harold C. Fritts

Climatic reconstructions have been derived from 65 tree-ring chronologies from western North America for 77 temperature stations and 96 precipitation stations throughout the United States and southwestern Canada. The seasonal and annual temperature and precipitation reconstructions are on a computer disk with this program, available from The University of Arizona Press or from the National Geophysical Data Center, 325 Broadway, Boulder, CO 80303, USA.

The tree-ring data were calibrated with climatic data for 1901–1963, verified against all available climatic data before 1901, and the results of the best two to three reconstructions (from models of different structure) were combined for the final reconstructions. The reconstructions included yearly estimates for 1602–1961 of the climate for winter (December–February), spring (March–June), summer (July–August), and autumn (September–November). These seasons were combined into annual (December–November) values.

Program DIFMAP was designed to (1) read either the seasonal or annual reconstructions for all grid points, (2) select reconstructions from any combination of years, (3) average the selected data, (4) make significance tests, (5) contour the data, and (6) make hard copies of the mapped patterns. In addition, an option was added that subtracts two maps, calculates the significance, and plots the difference. Up to ten maps can be created in one pass, and nine differences can be calculated from a single base period. If the base data are specified in the current run (i.e., not read from a file), the first set of years that is entered becomes the base data, and this map will be subtracted from all of the remaining sets of years that are specified.

The reconstructions are mapped for the entire United States, even though all reconstructions for the eastern states and many for the central states appear to be unreliable. It is recommended that reconstructions for a number of years be averaged together and evaluated using this program so that the errors of the individual estimates are minimized. For example, to evaluate the climate following volcanic eruptions, select a number of eruptions, average the climates lagging 1, 2, and 3 years behind the eruptions, and examine the mean map for each lag. If maps of years are examined individually, the uncertainty is large, and it is difficult to see relationships.

Program DIFMAP was originally a batch program written in FORTRAN IV for the Control Data Corporation Cyber System by Robert G. Lofgren and Kathy Sakai in March 1979. We have modified this program extensively for interactive mode on the VAX computer and compiled it to run on DOS using RYAN-McFARLAND/FORTRAN 1.1 and 2.4.

A. Input and Output Files

The program disk includes the following types of files:

Executable program: DIFMAP   EXE
Input data (i.e., annual temperature files):
    Reconstructions   TANRE   DAT
    Identification code   TANID   DAT
    Means   TANMD   DAT
    Standard deviations   TANST   DAT
Grid locations: TAPE71   DAT
Batch file: MAPCL   BAT
Fortran source code: AVERAGE.FOR, CINMIDP.FOR,
    COMPT.FOR, CONTUR2.FOR, CONVERT.FOR, DIFMAP.FOR,
    DAT1.FOR, INPUT1.FOR, INPUT2.FOR, INPUT3.FOR,
    MESS1.FOR, MESS2.FOR, PRINTMP.FOR, SELCINT.FOR,
    SIGT95.FOR, and STOPJOB.FOR.

Tape 71 includes the boundary and grid locations used in the mapping. The four data files are ABCID.DAT, ABCMD.DAT, ABCST.DAT, and ABCRE.DAT, where A is "T" for temperature, "R" for precipitation, and "P" for pressure (the pressure reconstruction maps use a different program that is in preparation); and B and C are the first two characters of the seasons or annual. The first output file includes ASCII characters and control codes for printing the maps under DOS. Control codes are produced for five types of printers. The earliest version of the program wrote files only for a printer with a carriage 14 inches (33.6 cm) wide. Options were added that reduce the image to 8.5 × 11-inch paper on Okidata ML 192, Diconix 150+, and Hewlett Packard Laserjet printers. Another option allows the entering of codes for other printers. The second output file includes a title and data for input to a DISPLAY plotting program, called HCFMAP, written by Robert G. Lofgren, that is maintained by the Laboratory of Tree-Ring Research on the VAX computer.

B. Running the Program

Start the program by entering "MAPCL"; this command calls a batch file that erases old output files before running the program. The DIFMAP program prompts and asks for the information needed to map the data. A default answer is also given in parentheses. The required information is entered on the keyboard, or a simple "return" (pressing the Enter key) selects the default answer.

The computer displays the following prompts (with the default answers shown in parentheses).

1. "Enter the name of the output file for the maps. (DIFMAP.OUT)"
   If you wish to write this file to another drive, precede the name by the drive name followed by a colon, i.e., "A:DIFMAP.OUT."

2. "Enter the name of the output file for the data in the maps. (DISPLAY.OUT)"
   This is the file for the DISPLAY program.

3. "Enter the number of your printer.
   (1) OKIDATA ML 192
   (2) DICONIX 150 +
   (3) LASERJET
   (4) PRINTER WITH WIDE CARRIAGE
   (5) OTHER PRINTERS (3)"
       This entry was described above.

   If "5" is entered, the following information is needed.

4. "(1) Read the print command from an existing file.
   (2) Enter the print command from the keyboard. (1)"
       If no commands have been entered on a previous run, select "2."

5. "Enter the file name with the printing command. (PRNSET.FIL)"
   The file name may include a drive and a colon. This file is saved and can be read as long as its contents have not been replaced by another code. It contains two lines of code.

6. "Enter the decimal number(s) in integer format I3 for printing 17 (16.67) characters per inch."

Refer to the character-width section of your printer manual. If only the ASCII codes are listed, convert them into decimal codes and enter them in this program.

7. "Enter the decimal numbers in integer format I3 for printing 13 (12) lines per inch."

Refer to the line-spacing section of the printer manual. This code is often composed of several decimal numbers. Enter the code for 13 (12) lines per inch or lines of 0.077 (0.083) inch.

The following information is used to select the input data.

8. "Enter the season:
   AN—Annual
   WI—Winter
   SP—Spring
   SU—Summer
   AU—Autumn (AN)"

9. "Enter the climate variable:
   T—Temperature
   R—Precipitation (T)"

10. "Enter the drive name: (C)"
    This is the drive with the reconstruction files.

For temperature, the following choices are offered:

11. "Enter the parameter to be mapped.
    F—Departures from the mean in degrees F
    C—Departures from the mean in degrees C
    N—Normalized values
    U—Untransformed data in degrees F
    X—Untransformed data in degrees C (C)"

For precipitation, the following choices are offered:

12. "Enter the parameter to be mapped.
    I—Departure from the mean in inches
    M—Departure from the mean in mm
    N—Normalized values

U—Untransformed values in inches
X—Untransformed values in mm
P—Percentage of calibration mean (P)"

The reconstructions are stored as normalized values, which are calculated as

$$Z_i = (X_i - X)/SD$$

where X_i is the original data at year i, and X and SD are the mean and standard deviation of the calibration period, respectively. Departures are calculated as

$$D_i = Z_i \times SD = X_i - X$$

and the untransformed data are

$$X_i = Z_i \times SD + X.$$

The percentage of calibration mean for precipitation data is

$$P_i = (D_i/X) * 100.$$

13. "Do you want a difference map? (Y)"
   This option calculates the differences between two maps. The difference involves a subtraction of A minus B, where B is defined as the BASE data. For a "Y" answer, the following information is requested:

14. "If you have saved a file in an earlier run, you can read that file for a BASE map. Do you want to read a file? If you do not have this file saved on disk, your answer must be 'N.' (N)"
   Prompts 15 and 16 follow a "Y" answer.

15. "Enter the local file name (DIFBAS.DAT)"

16. "Do you want to map the BASE data? (N)"
   Enter "N" if the BASE data have already been mapped in an earlier run and you want to avoid duplication. Usually, the BASE data are the first data to be selected and mapped.

17. "Maps of the differences are always mapped. Do you want to map the original data of the difference period? (N)" If the answer to this question is "N," the maps for A (defined above) will not be printed.

18. "How many difference maps? (1)"
    After an "N" answer to question 13, the following prompt is displayed:

19. "How many maps? (2)"

20. "The original calibration statistics are:
    1. The mean of the climatic data over the calibration period
    2. Standard deviation of the climatic data over the calibration period
    3. Standard deviation of the residuals
    4. Reduction-of-error statistic
    5. Calibrated variance, and
    6. Mean residuals
    Do you want to map them? (N)"
    The first two statistics describe the entire set of original climatic data. The last four statistics describe the entire set of reconstructions (see App. 1). These maps vary for each season and variable.

21. "Do you want to map the *t* values? (N)"

    A "Y" will produce additional maps of the sample *t*-statistic for each data point. With "N," the decimal point for all mean values with a significant *t*-statistic will be replaced by an asterisk. The 0.95 significance level for testing the *t*-statistic is a function of the numbers of degrees of freedom.
    The significance of a mean from zero is tested with a *t* statistic, where

$$t = (X - u)/\sqrt{[(SD^2 + SR^2)/N]}$$

and X and u are the sample and population means, respectively; SD^2 is the variance of the reconstructed data; SR^2 is the variance of the residuals; and N is the number of observations (years) in the sample.
    The significance of a difference between two means is

$$t = [(Xa - Xb) - (ua - ub)]/\sqrt{\{[(SRa^2 + SDa^2)*(Na - 1)}$$

$$+ (SRb^2 + SDb^2)*(Nb - 1)]*[(1/Na) + (1/Nb)]/(Na + Nb - 2)\}}$$

where Xa and Xb are the sample means, SDa and SDb the sample standard deviations, and SRa and SRb the residual variance for the samples; ua and ub are the population means, and Na and Nb are the number of years in the two samples. This test assumes that the two samples have the same population variance. One should note that this test differs from the usual t test in that the residual variance is added to the denominator to correct for the error of the reconstruction. Thus, areas of poor reconstruction are least likely to be significant.

22. "Use previously selected years from file READ.DAT? (N)"

The answers to items 23–25 are stored in a file which can be read at a later time. Enter "Y" to read this file instead of answering items 23 and 24. You will be allowed to change the title.

The program then reads the desired reconstructions into memory and requests the following information:

23. "Each map can be made from a number of elements. Each element is either a single year or a series of continuous years. The total number of elements can be 50. Enter the number of element(s) in the *1st* map, which is a xxxxxx map (1)"

"xxxxxx" may be BASE, AVERAGE, or DIFFERENCE, depending on the type of map, and the underlined characters (1st, 2nd, etc.) indicate which map in the sequence is being generated.

Each element represents a continuous period of one or more years. This feature provides maximum flexibility of input. For example, to obtain an average of years 1820–1825, 1833, 1900, and 1930–60, one enters "4," for there are four separate elements. The following message is then displayed.

24. "Two parameters define an element. They are the FIRST YEAR and the NUMBER OF YEARS. If NUMBER OF YEARS is blank, it will be set to 1. A 999 entered as YEARS will terminate and start the entry over again. Separate each entry with a space or comma. Enter FIRST YEAR and NUMBER OF YEARS in the 1st element. (1930 30)"

In the example above the first entry would be "1820 6." The next three entries would be "1833," "1900," and "1930 31." The program would average the reconstructions from these 39 years for each data point.

25. "Enter the title for this map [up to 40 characters in length]."

The 40-character title will be printed in the legend of the map. In addition, the type of map, the season, the data units, and years selected will be on the legend of each map.

26. "Do you want to save these data for a base period in a future DIFMAP run? Only 3 files can be saved in one run. (N)"
Prompt "27" follows a "Y" answer.

27. "Name the output file for saving these data. (DIFBAS.DAT)"
Use different names. If an old name is reassigned, an error can result.

28. "Do you wish to see the maps displayed on the screen? (Y)"
This is a feature added by Christopher Lambrecht, National Geophysical Data Center, Boulder, Colorado, which generates screen graphics using EGA resolution or better. If you do not have an EGA card or wish not to take the time to generate the screen map, answer "N."
The program then calculates the map and writes the maximum, minimum, and midpoint of the data to the screen to help in assigning contour intervals.

29. "The default mid-point corresponding to the $+/-$ contour line is x.xx. Do you want to change it? (N)"
We recommend answering "Y."

30. "Enter the mid-point to correspond to the $+/-$ contour line."
Enter 0.0 to ensure that the zero contour interval will fall exactly between "$+$" and "$-$" on the printed map making this interval easy to distinguish.

31. "The default contour interval is x.xx. Do you want to change it? (N)"
If the default interval is not a convenient rounded number, we recommend answering "Y."

32. "Enter the contour interval or 99 to abort."
Round the interval off by increasing it to the next higher value. A "99" entered at this point will abort the entry and return the program to item 29.

33. "Do you want more maps using the same periods or using the same climatic data? (N)"
This option allows one to save the instructions or data and make additional maps, a useful time saver if the instructions are complicated. With a "Y" answer the program returns to the beginning and asks for more input until you answer "N" to this question.

34. "1. New input data with the same periods.
    2.   The same input data with new periods. (2)"

If option "1" is chosen, the program control returns to the beginning; bypasses questions 1–3, 13, 19, 22, and 23; reads new climatic reconstructions; and calculates the statistics for the years that have already been selected. These data are displayed on the screen, and an opportunity is given to change the titles. If a difference map is requested under this option, the base data must be specified (as they cannot be read from a local file).

If option "2" is chosen, the program saves the current reconstructions, bypasses questions 1–10 and 20, and repeats questions 23 and 24, asking for new yearly data.

On termination of the program, and a "N" answer in item 33 terminates the following message is displayed:

"The output files are:
1. DIFMAP.OUT—FILE WITH MAP
2. DISPLAY.OUT—DATA FILE
3. READ.DAT—SELECTED YEARS FILE
Enter 'PRINT DIFMAP.OUT' to print the maps.
JOB COMPLETED"

Compiler Note

A problem was encountered in compiling subroutine AVERAGE because of a large array. It was resolved by using compiling option /b of RM-FORTRAN, which assumes that array sizes are larger than 65281 bytes.

Subroutine COMT does not compile under certain compilers. To correct this problem, the value of variable TENE75, which is used to check for large numbers, can be reduced.

Literature Cited

Abbe, C. 1893. The meteorological work of the U.S. Signal Service, 1870 to 1891. *Part 4 of Department of Agriculture, Weather Bureau Bulletin* 11:232–85.

Anderson, R. L., D. M. Allen, and F. B. Cady. 1972. Selection of predictor variables in linear multiple regression. In *Statistical Papers in Honor of George W. Snedecor*, ed. T. A. Bancroft. Ames: Iowa State University Press.

Anderson, R. Y., W. E. Dean, J. P. Bradbury, and D. Love. 1985. Meromictic lakes and varved lake sediments from North America. *U.S. Geological Survey, Bulletin*, 1607.

Andrade, E. R., Jr., and W. D. Sellers. 1989. El Niño and its effect on precipitation in Arizona and western New Mexico. *Journal of Climatology* 8:403–10.

Baron, W. R. 1980. Tempests, freshets and mackerel skies: Climatological data from diaries using content analysis. Ph.D. dissertation, University of Maine, Orno.

Barron, E. 1985. Climate models: Applications for the pre-Pleistocene. In *Paleoclimate Analysis and Modeling*, ed. A. D. Hecht, pp. 397–421. New York: John Wiley and Sons.

Barron, E., C. G. A. Harrison, J. L. Sloan II, and W. W. Hay. 1981. Paleogeography, 180 million years ago to the present. *Eclogae Geologicae Helvetiae* 74:443–70.

Barron, E., S. L. Thompson, and S. H. Schneider. 1981. An ice-free Cretaceous? Results from climate model simulations. *Science* 212:501–8.

Barry, R. G., and A. H. Perry. 1973. *Synoptic climatology methods and applications*. London: Methuen.

Bates, J. M. and C. W. J. Granger. 1969. The combination of forecasts. *Operational Research Quarterly* 20: 451–68.

Baumgartner, T. R., J. M. Michaelsen, L. G. Thompson, G. T. Shen, A. Soutar, and R. E. Casey. 1989. The recording of interannual climatic change by high-resolution natural systems: Tree-rings, coral bands, glacial ice layers, and marine varves. In *Aspects of climate variability in the Pacific and the western Americas*, ed. D. H. Peterson. *Geophysical Monograph* 55:1–14.

Bennett, R. J. 1979. *Spatial time series: Analysis-forecasting-control*. London: Pion.

Berger, A. 1979. Spectrum of climatic variations and their causal mechanisms. *Geophysical Surveys* 3:351–402.

Beyer, W. H., ed. 1968. *Handbook of tables for probability and statistics*. 2d ed. Cleveland: Chemical Rubber Co.

Bickel, P. J., and K. Doksum. 1977. *Mathematical statistics*. San Francisco: Holden-Day.

Blackman, R. B., and J. W. Tukey. 1958. *The measurement of power spectra*. New York: Dover.

Blasing, T. J. 1975. Methods for analyzing climatic variations in the North Pacific sector and western North America for the last few centuries. Ph.D. dissertation, University of Wisconsin, Madison.

———. 1978. Time series and multivariate analysis in paleoclimatology. In *Time series and ecological processes*, ed. H. H. Shugart, Jr., pp. 212–26. SIAM-SIMS Conference Series no. 5. Philadelphia: Society for Industrial and Applied Mathematics.

Blasing, T. J., and H. C. Fritts. 1975. Past climate of Alaska and northwestern Canada as reconstructed from tree rings. In *Climate of the Arctic: Proceedings of the 24th Alaska Scientific Conference* (August 15–17, 1973, Fairbanks), ed. G. Weller and S. A. Bowling, pp. 48–58. Fairbanks: Geophysical Institute, University of Alaska.

———. 1976. Reconstructing past climatic anomalies in the North Pacific and western North America from tree-ring data. *Quaternary Research* 6(4): 563–79.

Blasing, T. J. and G. R. Lofgren. 1980. Seasonal climatic anomaly types for the North Pacific sector and western North America. *Monthly Weather Review* 108:700–719.

Boden, T. A., P. Kanciruk and M. P. Farrell, eds. 1990. *Trends '90: A compendium of data on global change*. ORNL/CDIAC-36. Carbon Dioxide Information Analysis Center, Environmental Sciences Division, Oak Ridge National Laboratory, Oak Ridge, Tennessee. 267 pp.

Box, G. E. P., and G. H. Jenkins. 1976. *Time series analysis, forecasting, and control*. Rev. ed., San Francisco: Holden-Day.

Bradley, R. S. 1976a. *Precipitation history of the Rocky Mountain states*. Boulder: Westview Press.

———. 1976b. Seasonal precipitation fluctuations in the western United States during the late nineteenth century. *Monthly Weather Review* 104:501–12.

———. 1980. Secular fluctuations of temperature in the Rocky Mountain states and comparison with precipitation fluctuations. *Monthly Weather Review* 108:874–885.

———. 1985. *Quaternary paleoclimatology*. Boston: Allen and Unwin.

———. 1988. The explosive volcanic eruption signal in Northern Hemisphere continental temperature records. *Climatic Change* 12:221–43.

Bradley, R. S., H. F. Diaz, J. K. Eischeid, P. D. Jones, P. M. Kelly, and C. M. Goodess. 1987. Precipitation fluctuations over Northern Hemisphere land areas since the mid-19th century. *Science* 237:171–75.

Briffa, K. R., P. D. Jones, J. R. Pilcher, and M. K. Hughes. 1988. Reconstructing summer temperatures in northern Fennoscandinavia back to 1700 A.D.

using tree-ring data from Scots pine. *Arctic and Alpine Research* 20: 385–94.

Briffa, K. R., P. D. Jones, and F. H. Schweingruber. 1988. Summer temperature patterns over Europe: A reconstruction to 1750 A.D. based on maximum latewood density indices of conifers. *Quaternary Research*, 30:36–52.

Brubaker, L. B. 1980. Spatial patterns of tree-growth anomalies in the Pacific Northwest. *Ecology* 61: 798–807.

Brubaker, L. B., and E. R. Cook. 1983. Tree-ring studies of Holocene environments. In *Late Quaternary environments of the United States*, vol. 2, *The Holocene*, ed. H. E. Wright, Jr., pp. 222–35. Minneapolis: University of Minnesota Press.

Bryson, R. A. 1974. A perspective on climatic change. *Science* 184:753–60.

———. 1985. On climatic analogs in paleoclimatic reconstruction. *Quaternary Research* 23:275–86.

Bryson, R. A., and J. A. Dutton. 1961. Some aspects of the variance spectra of tree rings and varves. *Annals of the New York Academy of Science* 95:580–604.

Bryson, R. A., and F. K. Hare, eds. 1974. *Climates of North America*. Vol. 11 of *World survey of climatology*, ed. H. E. Landsberg. Amsterdam: Elsevier.

Bryson, R. A., and J. F. Lahey. 1958. *The march of the seasons*. ARCRC-TR-58-223. Final Report AF 19(604)992. Department of Meteorology, University of Wisconsin, Madison.

Budyko, M. I. 1969. The effect of solar radiation variations on the climate. *Tellus* 21:611–61.

———. 1974. *Climate and life*. London: Academic Press.

Burbank, D. W. 1981. A chronology of late Holocene glacier fluctuations on Mount Rainier, Washington. *Arctic and Alpine Research* 13:369–86.

Catchpole, A. J. W. 1985. Evidence from Hudson Bay region of severe cold in the summer of 1816. In *Climatic change in Canada 5*, ed. C. R. Harington, pp. 121–46. *Syllogeus* no. 55, National Museums of Canada, Ottawa.

Catchpole, A. J. W., and M. Faurer. 1983. Summer sea ice severity in Hudson Strait, 1751–1870. *Climatic Change* 5:115–39.

Catchpole, A. J. W., D. W. Moodie, and B. Kaye. 1970. Content analysis: A method for the identification of dates of first freezing and first breaking from descriptive accounts. *Professional Geographer* 22:252–57.

Clark, D. 1975. Understanding canonical correlation analysis. *Concepts and techniques in modern geography* 3. University of East Anglia, Norwich: Geographical Abstracts.

Cleaveland, M. K. 1986. Climatic response of densitometric properties in semiarid site tree rings. *Tree Ring Bulletin* 46:13–29.

CLIMAP. 1976. The surface of the Ice-Age. *Science* 191:1131–37.

———. 1981. Seasonal reconstruction of earth's surface at the last glacial maximum. *Geological Society of America, Map and Chart Series MC-36*.

Cook, E. R. 1985. A time-series analysis approach to tree-ring standardization. Ph.D. dissertation, University of Arizona, Tucson.

———. 1987. The decomposition of tree-ring series for environmental studies. *Tree-Ring Bulletin* 47:37–59.

Cook, E. R., K. R. Briffa, and P. D. Jones. 1988. Spatial regression methods for dendroclimatology: A review and comparison of two techniques. Part 1: The theory. Research report submitted to the Scientific Affairs Division, North Atlantic Treaty Organization, Brussels, Belgium.

Cook, E. R., and G. C. Jacoby, Jr. 1977. Tree-ring-drought relationships in the Hudson Valley, New York, *Science* 198:399–401.

————. 1979. Evidence for quasi-periodic July drought in the Hudson Valley, New York. *Nature* 282:390–92.

————. 1983. Potomac River streamflow since 1730 as reconstructed by tree rings. *Journal of Climate and Applied Meteorology* 22:1659–74.

Cook, E. R., and P. Mayes. 1985. Decadal-scale patterns of climatic change over eastern North America inferred from tree rings. *Book of abstracts and reports from the Conference on Abrupt Climatic Change* (Biviers, France, October 16–22). SIO Reference Series 86-8.

Cook, E. R., D. W. Stahle, and M. K. Cleaveland. In press. Dendroclimatic evidence from the eastern United States. In *Climate since A.D. 1500*, ed. R. S. Bradley and P. D. Jones. London: Unwin Hyman.

Cropper, J. P. 1982. Climate reconstructions from tree rings: Comment. In *Climate from tree rings*, ed. M. K. Hughes, P. M. Kelly, J. R. Pilcher, and V. C. LaMarche, Jr., pp. 65–67. Cambridge: Cambridge University Press.

————. 1984. Multicollinearity within selected western North American temperature and precipitation data sets. *Tree-Ring Bulletin* 44:29–37.

————. 1985. Tree-ring response functions: An evaluation by means of simulations. Ph.D. dissertation, University of Arizona, Tucson.

Cropper, J. P., and H. C. Fritts. 1982. Density of tree-ring grids in western North America. *Tree-Ring Bulletin* 42:3–9.

————. 1985. *A 360-year temperature and precipitation record for the Pasco Basin derived from tree-ring data*. Final report to Pacific Northwest Laboratories, Battelle Memorial Institute, Contract No. B-G5323-A-E. Hanford, Washington.

Dansgaard, W., S. J. Johnsen, H. B. Clausen, and C. C. Langway, Jr. 1971. Climatic record revealed by the Camp Century ice core. In *The Late Cenozoic glacial ages*, ed. K. Turekian, pp. 37–56. New Haven: Yale University Press.

Daultrey, S. 1976. Principal components analysis. *Concepts and techniques in modern geography* 8. University of East Anglia, Norwich: Geographical Abstracts.

Dean, J. S. 1978. *Tree-ring data in archaeology*. University of Utah Anthropology Papers 96. Salt Lake City: University of Utah Press.

Denton, G. H., and S. C. Porter. 1970. Neoglaciation. *Scientific American* 222:101–10.

DeVries, T. J. 1988. Paleoecology workshop. February 15–17, 1988, Boston. Report prepared for NOAA and NSF, 79 pp.

DeWitt, E. 1978. *Temperature and precipitation station selection*. Technical Note 3, Northern Climatic Reconstruction Group. Tucson: Laboratory of Tree-Ring Research, University of Arizona.

DeWitt, E., and M. Ames, eds. 1978. *Tree-ring chronologies of eastern North America. Vol. 1*. Chronology Series VI. Tucson: Laboratory of Tree-Ring Research, University of Arizona.

Diaz, H. F. 1983. Some aspects of major dry and wet periods in the contiguous United States, 1895–1981. *Journal of Climate and Applied Meteorology* 22:3–16.

_____. 1986. An analysis of twentieth-century climatic fluctuations in northern North America. *Journal of Climate and Applied Meteorology* 25:1625–57.

Diaz, H. F., and D. C. Fulbright. 1981. Eigenvector analysis of seasonal temperature, precipitation and synoptic-scale system frequency over the contiguous United States: Part 1, Winter. *Monthly Weather Review* 109:1267–84.

Diaz, H. F., and J. Namias. 1983. Associations between anomalies of temperature and precipitation in the United States and western Northern Hemisphere: 700 mb height profiles. *Climate and Applied Meteorology* 22:352–63.

Dickenson, R. E. 1989. Uncertainties of estimates of climatic change: A review. *Climatic Change* 15:5–13.

Draper, N. R., I. Guttman, and H. Kanemasu. 1971. The distribution of certain regression statistics. *Biometrika* 58: 295–98.

Dunn, O. J., and V. A. Clark. 1974. *Applied statistics: Analysis of variance and regression*. New York: John Wiley and Sons.

Dzerdzeevskii, B. L. 1968. *Circulation mechanisms in the atmosphere of the Northern Hemisphere in the twentieth century*. Moscow: Institute of Geography, Soviet Academy of Sciences.

Earth System Science Committee. 1988. *Earth System Science: A closer view*. Report of the Earth System Science Committee, NASA Advisory Council, Washington, D.C.

Eddy, J. A. 1976. The Maunder minimum. *Science* 192:1189–1202.

_____. 1977. Climate and the changing sun. *Climatic Change* 1:173–90.

Eddy, J. A., R. L. Gilliland, and D. V. Hoyt. 1982. Changes in the solar constant and climatic effects. *Nature* 300:689–93.

Fritts, H. C. 1971. Dendroclimatology and dendroecology. *Quaternary Research* 1: 419–49.

_____. 1974. Relationships of ring widths in arid-site conifers to variations in monthly temperature and precipitation. *Ecological Monograph* 44: 411–40.

_____. 1976. *Tree rings and climate*. London: Academic Press. Reprinted in 1987 in *Methods of dendrochronology*, vols. 2 and 3, ed. L. Kairiukstis, Z. Bednarz, and E. Felikstik. Warsaw: International Institute for Applied Systems Analysis and the Polish Academy of Sciences.

_____. 1977. Some quantitative methods for calibrating ring widths with variables of climate. In *Dendrochronologie und Postglaziale Klimatschwankungen in Europa*, ed. B. Frenzel. *Erdwissenschaftliche Forschung* 13:147–50. Wiesbaden: Franz Steiner Verlag GMBH. Germany.

_____. 1978. Tree rings, a record of seasonal variations in past climate. *Die Naturwissenschaften* 65:48–56.

————. 1981. Statistical climatic reconstructions from tree-ring widths. In *Climatic variations and variability: Facts and theories*, ed. A. Berger, pp. 135–53. Dordrecht, Netherlands: Reidel.

————. 1982. An overview of dendroclimatic techniques, procedures, and prospects. In *Climate from tree rings*, ed. M. K. Hughes, P. M. Kelly, J. R. Pilcher, and V.C. LaMarche, Jr., pp. 191–97. Cambridge: Cambridge University Press.

————. 1983. Tree-ring dating and reconstructed variations in central Plains climate. In *Man and the changing environments in the Great Plains*, ed. W. W. Caldwell, C. B. Schultz, and T. M. Stout, pp. 37–41. *Transactions of the Nebraska Academy of Sciences* 9.

————. 1984. Discussion of "Physical limitations of water resources," by J. Bredehoeft. In *Water scarcity: Impacts on western agriculture*, ed. E. A. Engelbert and A. F. Scheuring, pp. 44–48. Berkeley: University of California Press.

————. 1986. Historical changes in forest response to climatic variations and other factors deduced from tree rings. In *Climate change*, ed. J. G. Titus, pp. 39–58. Vol. 3 of *Effects of changes in stratospheric ozone and global climate*. Washington, D.C.: U.S. Environmental Protection Agency.

————. 1987. *Climatic regimes of the Pacific sector and adjacent continents since 1600: A synoptic description and comparison of independent climate proxy records*. Final Report, Project ATM-8319848. Climate Dynamics Program, National Science Foundation, Washington, D.C.

Fritts, H. C., and T. J. Blasing. 1974. Tree-ring analysis and its potential contribution to the mapping of past climates. In *Proceedings of the International CLIMAP Conference, May 17–22, 1973, Norwich, U.K., Collected Abstracts. Climatic Research Unit Research Publication* 2:17–20. University of East Anglia, Norwich.

Fritts, H. C., T. J. Blasing, B. P. Hayden, and J. E. Kutzbach. 1971. Multivariate techniques for specifying tree-growth and climate relationships and for reconstructing anomalies in paleoclimate. *Journal of Applied Meteorology* 10:845–64.

Fritts, H. C., T. J. Blasing, and G. R. Lofgren. 1977. Climatic variations reconstructed from tree rings. In *Man and nature—Makers of climatic variation*, Compendium of Papers, Centennial Academic Assemblies, College of Geosciences, Texas A and M University, College Station.

Fritts, H. C., E. DeWitt, G. A. Gordon, J. H. Hunt, and G. R. Lofgren. 1979. *Estimating long-term statistics for annual precipitation for six regions of the U.S. from tree-ring data*. Technical Report no. UCRI 15162 to Lawrence Livermore Laboratory, University of California.

Fritts, H. C., J. Guiot, G. A. Gordon, and F. Schweingruber. 1990. Methods of calibration, verification and reconstruction. In *Methods of Dendrochronology Applications in the environmental sciences*, ed. E. Cook and L. Kairiukstis. pp. 163–217. Dordrecht, Netherlands: Kluwer Academic Publishers.

Fritts, H. C., and G. R. Lofgren. 1978. *Patterns of climatic change revealed through dendroclimatology.* U.S. Army Coastal Engineering Research Center, Belvoir, Virginia.

Fritts, H. C., G. R. Lofgren, and G. A. Gordon. 1979. Variations in climate since 1602 as reconstructed from tree rings. *Quaternary Research* 12:18–46.

––––––. 1980. Past climate reconstructed from tree rings. *Journal of Interdisciplinary History* 10: 773–93.

––––––. 1981. Reconstructing seasonal to century time-scale variations in climate from tree-ring evidence. In *Climate and history*, ed. T. M. L. Wigley, M. J. Ingram, and G. Farmer, pp. 139–61. Cambridge: Cambridge University Press.

Fritts, H. C., and J. M. Lough. 1985. An estimate of average annual temperature variations for North America, 1602 to 1961. *Climatic Change* 7:203–24.

Fritts, H. C., J. E. Mosimann, and C. P. Bottorff. 1969. A revised computer program for standardizing tree-ring series. *Tree-Ring Bulletin* 29: 15–20.

Fritts, H. C., and X. Sha o. In press. Reconstructed climate from arid-site trees in western North America. In *Climate since A.D. 1500*, ed. R. S. Bradley and P. D. Jones. London: Unwin Hyman.

Fritts, H. C., and D. J. Shatz. 1975. Selecting and characterizing tree-ring chronologies for dendroclimatic analysis. *Tree-Ring Bulletin* 35:31–40.

Fritts, H. C., and T. W. Swetnam. 1989. Dendroecology: A tool for evaluating variations in past and present forest environments. *Advances in Ecological Research* 19:111–88.

Fritts, H. C., E. A. Vaganov, I. V. Sviderskaya, and A. V. Shashkin. 1991. Climatic variation and tree-ring structure in conifers: empirical and mechanistic models of tree-ring width, number of cells, cell size, cell-wall thickness and wood density. *Climate Research* 1:97–116.

Fritts, H. C., and Wu X. 1986. A comparison between response function analysis and other regression techniques. *Tree-Ring Bulletin* 46:31–46.

Garfinkel, H. L., and L. B. Brubaker. 1980. Modern climate—Tree-growth relationships and climatic reconstruction in subarctic Alaska. *Nature* 286:872–74.

Gates, W. L., and Y. Mintz. 1975. *Understanding climatic change: A program for action.* Report of the Panel on Climatic Variation of the U.S. Committee for GARP (Global Atmosphere Research Project), National Research Council, National Academy of Sciences, Washington, D.C.

Glahn, H. R. 1968. Canonical correlation and its relationship to discriminant analysis and multiple regression. *Journal of Atmospheric Science* 25: 23–31.

Gordon, G. A. 1980. *Verification tests for dendroclimatological reconstructions.* Technical Note 19, Northern Hemisphere Climate Reconstruction Group. Tucson: Laboratory of Tree-Ring Research, University of Arizona.

––––––. 1982. Verification of dendroclimatic reconstructions. In *Climate from tree rings*, ed. M. K. Hughes, P. M. Kelly, J. R. Pilcher, and V. C. LaMarche, Jr., pp. 58–61. Cambridge: Cambridge University Press.

Gordon, G. A., and S. K. LeDuc. 1981. Verification statistics for regression models. *Seventh Conference on Probability and Statistics in Atmospheric Sciences*, Session 8, pp. 129–133. American Meteorological Society, Boston.

Gordon, G. A., J. M. Lough, H. C. Fritts, and P. M. Kelly. 1985. Comparison of sea-level pressure reconstructions from western North American tree rings with a proxy record of winter severity in Japan. *Journal of Climate and Applied Meteorology* 24:1219–24.

Graumlich, L. J. 1985. Long-term records of temperature and precipitation in the Pacific Northwest derived from tree rings. Ph.D. dissertation, University of Washington, Seattle.

_____. 1991. Subalpine tree growth, climate, and increasing CO_2: an assessment of recent growth trends. *Ecology* 72:1–11.

Graumlich, L. J., and L. B. Brubaker. 1986. Reconstruction of annual temperatures (1590–1979) for Longmire, Washington, derived from tree-rings. *Quaternary Research* 25:223–34.

_____. 1987. Increasing net primary productivity in Washington (U.S.A.) forests during the last 1000 years. In *Proceedings of the International Symposium on Ecological Aspects of Tree-Ring Analysis* (August 1986, Tarrytown, New York), pp. 59–69. U.S. Department of Energy CONF-8608144.

Graybill, D. A. 1979a. *Program operating manual for RWLIST, INDEX and SUMAC*. Tucson: Laboratory of Tree-Ring Research, University of Arizona.

_____. 1979b. Revised computer programs for tree-ring research. *Tree-Ring Bulletin* 39:77–82.

_____. 1982. Chronology development and analysis. In *Climate from tree rings*, ed. M. K. Hughes, P. M. Kelly, J. R. Pilcher and V. C. LaMarche, Jr., pp. 21–30. Cambridge: Cambridge University Press.

_____. 1987. A network of high-elevation conifers in the western U.S. for detection of tree-ring growth response to increasing atmospheric carbon dioxide. *Proceedings of the International Symposium on Ecological Aspects of Tree-Ring Analysis* (August 1986, Tarrytown, New York), pp. 463–74. U.S. Department of Energy CONF-8608144.

Grove, J. M. 1988. *The Little Ice Age*. New York: Methuen.

Groveman, B. S., and H. E. Landsberg. 1979. Simulated Northern Hemisphere temperature departures, 1579–1880. *Geophysical Research Letters* 6:767–69.

Guiot, J. 1987. Reconstruction of seasonal temperatures in central Canada since A.D. 1700 and detection of the 18.6- and 22-year signals. *Climatic Change* 10:249–68.

Guiot, J., A. Berger, A. V. Munaut, and C. Till. 1982. Some new mathematical procedures in dendroclimatology with examples in Switzerland and Morocco. *Tree-Ring Bulletin* 42:33–48.

Hammer, C. U., H. B. Clausen, and W. Dansgaard. 1980. Greenland ice sheet evidence of post-glacial volcanism and its climatic impact. *Nature* 288:230–35.

Hansen, J., D. Rind, A. DelGenio, A. Lacis, S. Lebedeff, M. Prather, and R. Ruedy. 1988. Regional greenhouse climate effects. *Proceedings of the*

Second North American Conference on Preparing for Climatic Change (December 6–8, 1988). Washington, D.C.: Climate Institute.

Hays, J. D., J. Imbrie, and N. Shackelton. 1976. Variations in the earth's orbit: Pacemaker of the ice age. *Science* 194–1121:32.

Hecht, A. D., ed. 1985. *Paleoclimate analysis and modeling.* New York: John Wiley and Sons.

Heikkinen, O. 1984. Climatic changes during recent centuries as indicated by dendrochronological studies, Mount Baker, Washington, U.S.A. In *Climatic changes on a yearly to millennial basis,* ed. N. A. Morner and W. Karlen, pp. 353–61. Dordrecht, Netherlands: Reidel.

Hughes, M. K. 1987. Requirements for spatial and temporal coverage: Introduction. In *Methods of dendrochronology—I, Proceedings of the Task Force Meeting on Methodology of Dendrochronology: East/west approaches,* ed. L. Kairiukstis, Z. Bednarz, and E. Feliksik, pp. 107–15. Warsaw: International Institute of Applied Systems Analysis and Polish Academy of Science.

Hughes, M. K., P. M. Kelly, J. R. Pilcher, and V. C. LaMarche, Jr., eds. 1982. *Climate from tree rings.* Cambridge: Cambridge University Press.

Hughes, M. K., F. H. Schweingruber, D. Cartwright, and P. M. Kelly, 1984. July–August temperature at Edinburgh between 1721 and 1975 from tree-ring density and width data. *Nature* 308:341–44.

Imbrie, J., and J. Z. Imbrie. 1980. Modeling the climatic response to orbital variations. *Science* 207:943–53.

Jacoby, G. C., ed. 1980. *Proceedings of the International Meeting on Stable Isotopes in Tree-Ring Research* (New Paltz, New York). U.S. Department of Energy, Carbon Dioxide Effects Research and Assessment Program, Publication no. 12, CONF-790518, UC-11, Washington, D.C.

Jacoby, G. C., E. R. Cook, and L. D. Ulan. 1985. Reconstructed summer degree days in central Alaska and northwestern Canada since 1524. *Quaternary Research* 23:18–26.

Jenkins, G. M., and D. G. Watts. 1968. *Spectral analysis and its applications.* San Francisco: Holden-Day.

Johnson, C. M. 1980. Wintertime Arctic sea ice extremes and the simultaneous atmospheric circulation. *Monthly Weather Review* 108:1782–91.

Johnston, J. 1972. *Econometric methods.* 2d ed. New York: McGraw-Hill.

Jones, P. D. 1987. The early twentieth century Arctic high—Fact or fiction? *Climate Dynamics* 1:63–75.

Jones, P. D., T. M. L. Wigley, and P. M. Kelly. 1982. Variations in surface air temperatures: Part 1, Northern Hemisphere, 1881–1980. *Monthly Weather Review* 110:59–70.

Jones, P. D., T. M. L. Wigley, and P. B. Wright. 1986. Global temperature variations, 1861–1984. *Nature* 322:430–34.

Kellogg, W. W. 1987. Mankind's impact on climate: the evolution of an awareness. *Climatic Change* 10:113–136.

Kelly, P. M., P. D. Jones, C. B. Sear, B. S. G. Cherry, and R. K. Tavakol. 1982. Variations in surface air temperatures: Part 2, Arctic regions, 1881–1980. *Monthly Weather Review* 110:71–83.

Kelly, P. M., and C. B. Sear. 1984. Climatic impact of explosive volcanic eruptions. *Nature* 311:740–43.

Kennedy, E. A., and G. A. Gordon. 1980. *Characteristics of nineteenth-century climatic data used in verification of dendroclimatic reconstructions*. Technical Note 16, Northern Climatic Reconstruction Group. Tucson: Laboratory of Tree-Ring Research, University of Arizona.

Kienast, F., and R. J. Luxmoore. 1988. Tree-ring analysis and conifer growth responses to increased atmospheric CO_2 levels. *Oecologia* 76:487–95.

Kocharov, G. E. 1986. Cosmic ray archaeology, solar activity and supernova explosions. Preprint 1039. A.F.IOFFE Physico-Technical Institute, Academy of Sciences of the USSR. 31 pp.

Kohler, M. A. 1949. On the use of double-mass analysis for testing the consistency of meteorological records and for making required adjustments. *Bulletin of the American Meteorological Society* 30:188–89.

Kozlowski, T. T. 1971. *Cambial growth, root growth and reproductive growth*. Vol. 2 of *Growth and development of trees*. New York: Academic Press.

Kramer, P. J., and T. T. Kozlowski. 1979. *Physiology of woody plants*. New York: Academic Press.

Kutzbach, J. E. 1970. Large-scale features of monthly mean Northern Hemisphere anomaly maps of sea-level pressure. *Monthly Weather Review* 98:708–16.

Kutzbach, J. E., and P. J. Guetter. 1980. On the design of paleoenvironmental data networks for estimating large-scale circulation patterns. *Quaternary Research* 14:169–87.

Ladurie, E. L. 1971. *Times of feast, times of famine: A history of climate since the year 1000*. Garden City, New York: Doubleday.

LaMarche, V. C., Jr. 1973. Holocene climatic variations inferred from treeline fluctuations in the White Mountains, California. *Quaternary Research* 3:632–60.

_____. 1974. Frequency-dependent relationships between tree-ring series along an ecological gradient and some dendroclimatic implications. *Tree-Ring Bulletin* 34:1–20.

_____. 1978. Tree-ring evidence of past climatic variability. *Nature* 276:334–38.

_____. 1982. Sampling strategies. In *Climate from tree rings*, ed. M. K. Hughes, P. M. Kelly, J. R. Pilcher, and V. C. LaMarche, Jr., pp. 2–6. Cambridge: Cambridge University Press.

LaMarche, V. C., Jr., and H. C. Fritts. 1971. Tree rings, glacial advance, and climate in the Alps. *Zeitschrift fur Gletscherkunde und Glazialgeologie* 7: 125–32.

LaMarche, V.C., Jr., D. A. Graybill, H. C. Fritts, and M. R. Rose. 1984. Increasing atmospheric carbon dioxide: Tree-ring evidence for growth enhancement in natural vegetation. *Science* 225:1019–21.

LaMarche, V. C., Jr., and K. K. Hirschboeck. 1984. Frost rings in trees as records of major volcanic eruptions. *Nature* 307:121–26.

LaMarche, V. C., Jr., R. L. Holmes, P. W. Dunwiddie, and L. G. Drew. 1979a. *Tree-ring chronologies of the Southern Hemisphere. Vol. 1. Argentina.*

Chronology Series V. Tucson: Laboratory of Tree-Ring Research, University of Arizona.

——. 1979b. *Tree-ring chronologies of the Southern Hemisphere. Vol. 2. Chile.* Chronology Series V. Tucson: Laboratory of Tree-Ring Research, University of Arizona.

——. 1979c. *Tree-ring chronologies of the Southern Hemisphere. Vol. 4. Australia.* Chronology Series V. Tucson: Laboratory of Tree-Ring Research, University of Arizona.

LaMarche, V. C., Jr., and C. W. Stockton. 1974. Chronologies from temperature-sensitive bristlecone pines at upper treeline in western United States. *Tree-Ring Bulletin* 34:21–45.

Lamb, H. H. 1969. Climatic fluctuations. In *World survey of climatology. Vol. 2. General climatology*, ed. H. Flohn, pp. 173–249. Amsterdam: Elsevier.

——. 1970. Volcanic dust in the atmosphere; with a chronology and assessment of its meteorological significance. *Philosophical Transactions of the Royal Society of London* 226: 425–533.

——. 1972. *Climate: Present, past and future. Vol. 1. Fundamentals and climate now.* London: Methuen.

——. 1977. *Climate: Present, past and future. Vol. 2. Climatic history and the future.* London: Methuen.

Landsberg, H. E. 1967. Two centuries of New England climate. *Weatherwise* 20:52–57.

Landsberg, H. E., J. M. Mitchell, Jr., H. L. Crutcher, and F. T. Quinlan. 1963. Surface signs of the biennial atmospheric pulse. *Monthly Weather Review* 91:549–56.

Landsberg, H. E., C. S. Yu, and L. Huang. 1968. *Preliminary reconstruction of a long time series of climatic data for the eastern United States.* Technical Note BN-571, Institute of Fluid Dynamics. College Park: University of Maryland.

Larson, S. C. 1931. The shrinkage of the coefficient of multiple correlation. *Journal of Educational Psychology* 22:45–55.

Lawson, M. P. 1974. *The climate of the great American desert: Reconstruction of the climate of the western interior United States, 1800–1850.* University of Nebraska Studies, New Series 46. Lincoln: University of Nebraska.

Leavitt, S. W., and A. Long. 1988. Stable carbon isotope chronologies from trees in the southwestern United States. *Global Biogeochemical Cycles* 2:189–98.

Liverman, K. W., W. H. Terjung, J. T. Hayes, and L. O. Mearns. 1986. Climatic change and grain corn yields in the North American Great Plains. *Climatic Change* 9:327–47.

Livezey, R. E., and W. Y. Chen. 1983. Statistical field significance and its determination by Monte Carlo techniques. *Monthly Weather Review* 111:46–59.

Lofgren, G. R. 1978. *Comparison of the monthly sea-level pressure data sets from the National Center for Atmospheric Research and from the Laboratory of Tree-Ring Research.* Technical Note 1, Northern Hemisphere Climatic

Reconstruction Group. Tucson: Laboratory of Tree-Ring Research, University of Arizona.

Lofgren, G. R., and J. H. Hunt. 1982. Transfer functions. In *Climate from tree rings*, ed. M. K. Hughes, P. M. Kelly, J. R. Pilcher, and V. C. LaMarche, Jr., pp. 50–56. Cambridge: Cambridge University Press.

Lorenz, E. N. 1956. *Empirical orthogonal functions and statistical weather prediction*. M.I.T. Statistical Forecasting Project, Scientific Report no. 1. Contract no. AF(604)-1566.

_____. 1977. An experiment in nonlinear statistical weather forecasting. *Monthly Weather Review* 105:590–602.

Lough, J. M. In press. The climate of 1816 and 1811–1820 as reconstructed from western North American tree ring chronologies. In *The year without a summer? Climate in 1816* pp. 96–113. (Ottawa, Canada, June 25–28).

Lough, J. M., and D. J. Barnes. 1990. Possible relationships between environmental variables and skeletal density in a coral colony from the central Great Barrier Reef. *Journal of Experimental Marine Biology and Ecology* 134:221–41.

Lough, J. M., and H. C. Fritts. 1985. The Southern Oscillation and tree rings: 1600–1961. *Journal of Climate and Applied Meteorology* 24:952–66.

_____. 1987. An assessment of the possible effects of volcanic eruptions on North American climate using tree-ring data, 1602 to 1900 A.D. *Climatic Change* 10:219–39.

_____. 1990. Historical aspects of El Niño/Southern Oscillation—Information from tree rings. In *Global ecological consequences of the 1982–1983 El Niño-Southern Oscillation*, ed. P. W. Glynn. Elsevier Oceanography Series 52, pp. 285–321. New York and Amsterdam: Elsevier.

Lough, J. M., H. C. Fritts, and Wu X. 1987. Relationships between the climates of China and North America over the past four centuries: A comparison of proxy records. In *The climate of China and global climate: Proceedings of the Beijing International Symposium on Climate* (Oct.–Nov. 1984, Beijing, China), ed. Ye D., Fu C., Chao J., and M. Yoshino, pp. 89–105. Beijing: China Ocean Press; Berlin: Springer.

Ludlum, D. M. 1966. *The history of American weather: Early American winters. Vol. 1. 1604–1820*. Boston: American Meteorological Society.

_____. 1968. *The history of American Weather: Early American winters. vol. 2. 1821–1870*. Boston: American Meteorological Society.

McCarthy, P. J. 1976. The use of balance half-sample replication in cross-validation studies. *Journal of the American Statistical Association* 71:596–604.

Manley, G. 1974. Central England temperatures: Monthly means from 1659 to 1973. *Quarterly Journal of the Royal Meteorological Society* 100:389–405.

Meko, D. M. 1981. Applications of Box-Jenkins methods of time-series analysis to the reconstruction of drought from tree rings. Ph.D. dissertation, University of Arizona, Tucson.

_____. 1982. Drought history in the western Great Plains from tree rings. Proceedings of the International Symposium on Hydrometeorology, pp. 321–26. Technical Publication Series TPS-82-1, 321-26. Bethesda, Maryland: American Water Resources Association.

Meko, D. M., C. W. Stockton, and W. R. Boggess. 1980. A tree-ring reconstruction of drought in southern California. *Water Resources Bulletin* 16: 594–600.

Michaelsen, J., L. Haston, and F. W. Davis. 1987. Four hundred years of central California precipitation variability reconstructed from tree-rings. *Water Resources Bulletin* 23: 809–18.

Mitchell, J. M., Jr. 1976. An overview of climatic variability and its causal mechanisms. *Quaternary Research* 6:481–93.

Mitchell, J. M., Jr., B. Dzerdzeevskii, H. Flohn, W. L. Hofmeyr, H. H. Lamb, K. N. Rao, and C. C. Wallen. 1966. *Climatic change*. Technical Note 79. Geneva: World Meteorological Organization.

Monserud, R. A. 1986. Time-series analyses of tree-ring chronologies. *Forest Science* 32: 349–72.

Mosteller, F., and J. W. Tukey. 1968. Data analysis, including statistics. In *Handbook of social psychology*, vol. 2, ed. G. Lindzey and E. Aronson. Reading, Massachusetts: Addison-Wesley.

_____. 1977. *Data analysis and regression*. Reading, Massachusetts: Addison-Wesley.

Namias, J. 1982a. Sea surface temperature teleconnections in the North Pacific and related coastal phenomena. In *Preprint volume: First International Conference on Meteorology and Air/Sea Interactions of the Coastal Zone* (May 10–14, 1982), pp. 301–4. Boston: American Meteorological Society.

_____. 1982b. Anatomy of Great Plains protracted heat waves (especially the 1980 U.S. summer drought). *Monthly Weather Review* 110:824–38.

National Academy of Sciences. 1975. *Understanding climatic change*. Washington, D.C.: National Academy of Sciences.

O'Sullivan, P. E. 1983. Annually laminated lake sediments and the study of Quaternary environmental changes—a review. *Quaternary Science Reviews* 1:245–313.

Palmer, W. C. 1965. *Meteorological drought*. U.S. Weather Bureau Research Paper 45. Washington, D.C.: U.S. Government Printing Office.

Paltridge, G., and S. Woodruff. 1981. Changes in global surface temperature from 1880 to 1977 derived from historical records of sea surface temperature. *Monthly Weather Review* 109:2427–34.

Panofsky, H. A., and G. W. Brier. 1968. *Some applications of statistics to meteorology*. University Park: Pennsylvania State University.

Parker, M. L., and W. E. S. Henoch. 1971. The use of Engelmann spruce latewood density for dendrochronological purposes. *Canadian Journal of Forest Research* 1: 90–98.

Parry, M. L., and T. R. Carter. 1989. An assessment of the effects of climatic change on agriculture. *Climatic Change* 15:95–116.

Pittock, A. B. 1978. A critical look at long-term sun-weather relationships. *Reviews of Geophysics and Space Physics* 16:400–420.

_____. 1982. Climatic reconstructions from tree rings. In *Climate from tree rings*, ed. M. K. Hughes, P. M. Kelly, J. R. Pilcher, and V. C. LaMarche Jr., pp. 62–63. Cambridge: Cambridge University Press.

Pittock, A. B., and H. A. Nix. 1986. The effect of changing climate on Australian biomass production—a preliminary study. *Climatic Change* 8:243–255.

Porter, S. C. 1981. Recent glacier variations and volcanic eruptions. *Nature* 5811:139–42.

Porter, S. C., and G. H. Denton. 1967. Chronology of neoglaciation in the North American Cordillera. *American Journal of Science* 265:177–210.

Preisendorfer, R. W., and T. P. Barnett. 1977. Significance tests for empirical orthogonal functions. In *Preprint volume: Fifth Conference on Probability and Statistics* (November 15–18, 1977). Boston: American Meteorological Society.

Preisendorfer, R. W., F. W. Zwiers, and T. P. Barnett. 1981. *Foundations of principal component selection rules*. SIO Reference Series 81-4. San Diego, La Jolla: Scripps Institution of Oceanography, University of California.

Rampino, M. R., and S. Self. 1982. Historic eruptions of Tambora (1815), Krakatoa (1883) and Agung (1963), their stratospheric aerosols and climatic impact. *Quaternary Research* 18:127–63.

Rencher, A. C., and F. C. Pun. 1980. Inflation of R^2 in best subset regression. *Technometrics* 22: 49–53.

Robock, A. 1979. The "Little Ice Age": Northern Hemisphere average observations and model calculations. *Science* 206:1602–4.

Rose, M. 1983. Time domain characteristics of tree-ring chronologies and eigenvector amplitude series from western North America. Technical Note 25, Northern Hemisphere Climatic Reconstruction Group. Tucson: Laboratory of Tree-Ring Research, University of Arizona.

Rumbaugh, W. F. 1934. The effect of time of observation on mean temperature. *Monthly Weather Review* 62:375–76.

Saltzman, B. 1985. Paleoclimatic modeling. In *Paleoclimate analysis and modeling*, ed. A. D. Hecht, pp. 341–96. New York: John Wiley and Sons.

Schlesinger, M. E., and J. F. B. Mitchell. 1987. Climate model simulations of the equilibrium climatic response to increased carbon dioxide. *Reviews of Geophysics* 25:760–98.

Schneider, S. H. 1989. The greenhouse effect: Science and Policy. *Science* 243:771–81.

Schneider, S. H., and C. Mass. 1975. Volcanic dust, sunspots and temperature trends. *Science* 190:741–46.

Schonher, T., and S. E. Nicholson. 1989. The relationship between California rainfall and ENSO events. *Journal of Climatology* 2:1258–69.

Schulman, E. 1951. Tree-ring indices of rainfall, temperature, and river flow. In *Compendium of Meteorology*, pp. 1024–29. Boston: American Meteorological Society.

Schulman, E. 1956. *Dendroclimatic changes in semiarid America*. Tucson: University of Arizona Press.

Schweingruber, F. H. 1988. A new dendroclimatological network in western North America. *Dendrochronologia* 6:171–178.

Schweingruber, F. H. 1988. *Tree rings, basics and applications of dendrochronology*. Dordrecht, Netherlands: Reidel.

Schweingruber, F. H., O. U. Braker, and E. Schar. 1979. Dendroclimatic studies on conifers from central Europe and Great Britain. *Boreas* 8:427–52.

Schweingruber, F. H., K. R. Briffa, and P. D. Jones. 1991. Yearly maps of summer temperatures in Western Europe from A.D. 1750 to 1975 and Western North America from 1600 to 1982: Results of a radiodensitometrical study of tree rings. *Vegetatio* 92(1):5–71.

Scuderi, L. A. 1987. Late Holocene upper timberline variation in the southern Sierra Nevada. *Nature* 325:242–44.

Self, S., M. R. Rampino, and J. J. Barbera. 1981. The possible effects of large 19th- and 20th-century volcanic eruptions on zonal and hemispheric surface temperatures. *Journal of Volcanology and Geothermal Research* 11:41–60.

Stahle, D. W., and M. K. Cleaveland. 1988. Texas drought history reconstructed and analyzed from 1698 to 1980. *Journal of Climate* 1:59–74.

Stahle, D. W., M. K. Cleaveland, and J. G. Hehr. 1988. North Carolina climate changes reconstructed from tree rings: A.D. 372 to 1985. *Science* 240:1517–19.

Stockton, C. W., W. R. Boggess, and D. M. Meko. 1985. Climate and tree rings. In *Paleoclimate analysis and modeling*, ed. A. D. Hecht, pp. 71–161. New York: John Wiley and Sons.

Stockton, C. W., and H. C. Fritts. 1971. Conditional probability of occurrence for variations in climate based on width of annual tree rings in Arizona. *Tree-Ring Bulletin* 31:3–24.

Stockton, C. W., and G. C. Jacoby, Jr. 1976. *Long-term surface-water supply and streamflow trends in the upper Colorado River Basin based on tree-ring analysis*. Lake Powell Research Project Bulletin no. 18. Los Angeles: University of California, Los Angeles Institute of Geophysics and Planetary Physics.

Stockton, C. W., and D. M. Meko. 1975. A long-term history of drought occurrence in western United States as inferred from tree rings. *Weatherwise* 28: 244–49.

―――――. 1983. Drought recurrence in the Great Plains as reconstructed from long-term tree-ring records. *Journal of Climate and Applied Meteorology* 22:17–29.

Stockton, C. W., J. M. Mitchell, Jr., and D. M. Meko. 1983. A reappraisal of the 22-year drought cycle. In *Weather and climate responses to solar variations*, ed. B. M. McCormac, pp. 507–15. Boulder: Colorado Associated University Press.

Stokes, M. A., and T. L. Smiley. 1968. *An introduction to tree-ring dating*. Chicago: University of Chicago Press.

Stone, M. 1974. Cross-validatory choice and assessment of statistical predictions. *Journal of the Royal Statistical Society*, Series B, 36: 11–147.

Stuiver, M. 1978. Atmospheric carbon dioxide and carbon reservoir changes. *Science* 199:245–49.

_____. 1980. Solar variability and climatic change during the current millennium. *Nature* 286:868–71.

Stuiver, M., and P. D. Quay. 1980. Changes in atmospheric carbon-14 attributed to a variable sun. *Science* 207:11–19.

Tatsuoka, M. 1974. *Multivariate analysis: Techniques for educational and psychological research*. New York: John Wiley and Sons.

Thompson, L. G., E. Mosley-Thompson, J. F. Bolzan, and B. R. Koci. 1985. A 1500-year record of tropical precipitation in ice cores from the Quelccaya ice cap, Peru. *Science* 229:971–73.

Thompson, L. G., E. Mosley-Thompson, W. Dansgaard, and P. M. Grootes. 1986. The Little Ice Age as recorded in the stratigraphy of the tropical Quelccaya ice cap. *Science* 234:361–64.

Thompson, L. G., E. Mosley-Thompson, and B. Morales Arnao. 1984. El Niño-Southern Oscillation events recorded in the stratigraphy of the tropical Quelccaya ice cap, Peru. *Science* 226:50–53.

Trenberth, K. E., G. W. Branstator, and P. A. Arkin. 1988. Origins of the 1988 North American drought. *Science* 242:1640–45.

Trenberth, K. E., and D. A. Paolino, Jr. 1980. The Northern Hemisphere sea-level pressure data set: Trends, errors, and discontinuities. *Monthly Weather Review* 108:855–72.

U.S. Department of Commerce, Weather Bureau. 1950. *Local climatic summary with comparative data*. Washington, D.C.: U.S. Government Printing Office.

van Loon, H., and J. Williams. 1976. The connection between trends of mean temperature and circulation at the surface: Part 1, Winter. *Monthly Weather Review* 104:365–80.

_____. 1980. The association between latitudinal temperature gradient and eddy transport: Part 2, Relationships between sensible heat transport by stationary waves and wind, pressure and temperature in winter. *Monthly Weather Review* 108:604–14.

Villalba, R. 1990. Climatic fluctuations in northern Patagonia during the last 1000 years as inferred from tree-ring records. *Quaternary Research* 34:346–360.

Wahl, E. W. 1968. A comparison of the climate of the eastern United States during the 1830's with the current normals. *Monthly Weather Review* 96:73–82.

Wahl, E. W., and M. L. Lawson. 1970. The climate of the mid-19th century United States compared to the current normals. *Monthly Weather Review* 98:259–65.

Walsh, J. E., and A. Mostek. 1980. A quantitative analysis of meteorological anomaly patterns over the United States, 1900–1977. *Monthly Weather Review* 108:615–30.

Webb, T., III, and D. R. Clarke. 1977. Calibrating micropaleontological data in climatic terms: A critical review. *Annals of the New York Academy of Sciences* 288:93–118.

Wendland, W. M., and R. A. Bryson. 1981. Northern Hemisphere airstream regions. *Monthly Weather Review* 109:255–70.

Wherry, R. J. 1931. A new formula for predicting the shrinkage of the multiple correlation coefficient. *Annals of Mathematical Statistics* 2:440–57.

Wigley, T. M. L., K. R. Briffa, and P. D. Jones. 1984. On the average value of correlated time series, with applications in dendroclimatology and hydro-meteorology. *Journal of Climate and Applied Meteorology* 23:201–13.

Wigley, T. M. L., B. M. Gray, and P. M. Kelly. 1978. Climatic interpretation of delta 18 oxygen and delta deuterium in tree rings. *Nature* 271:92–93.

Wood, F. B. 1988a. Global alpine glacier trends, 1960s–1980s. *Arctic and Alpine Research* 20:404–13.

_____. 1988b. The need for systems research on global climate change. *Systems Research* 5: 225–40.

Wright, H. E., Jr. 1983. *Late Quaternary environments of the United States*. Vols. 1 and 2, Minneapolis: University of Minnesota Press.

Index